# ENVIRONMENTAL ETHICS

## Intercultural Perspectives

VIBS

Volume 205



a volume in
**Studies in Applied Ethics**
**SAE**
Gerhold K. Becker, Editor

# ENVIRONMENTAL ETHICS

## Intercultural Perspectives

Edited by
King-Tak Ip

Rodopi

Amsterdam - New York, NY 2009

Cover Design: Studio Pollmann

The paper on which this book is printed meets the requirements of "ISO 9706:1994, Information and documentation - Paper for documents - Requirements for permanence".

ISBN: 978-90-420-2595-0

Printed in the Netherlands

# CONTENTS

# EDITORIAL FOREWORD

On 2 February 2007, the United Nations Intergovernmental Panel on Climate Change released its report on global warming. In view of their findings, the world can no longer afford to ignore compelling evidence for ominous developments in the earth's climate. For the first time, a United Nations panel concluded that global warming is "unequivocal" and that human activity is the main driver, "very likely" causing most of the rise in temperatures since 1950. While the report's conclusions had to be somewhat softened due to pressure by some governments, the data could not be suppressed and speak loud and clear for themselves. Over the next fifty years, the world will see the displacement of millions of people due to rising sea levels, extreme water shortages affecting the lives of one in six of the world's population, droughts creating tens of millions of climate refugees, and the likely extinction of about forty percent of wild life, insect, and plant species.

The ecological crisis is not simply a political issue that we could leave to governments to handle but a threat to planet earth that concerns each one of us. It is also a moral challenge requiring us once more to accept responsibility for our actions and to conscientiously take into account their implications for future generations. This seems to imply attitudinal changes in our dealing with nature and its resources and a re-assessment of the vision of a "good life" that has traditionally given direction to our actions.

The present volume is a timely reminder that in this endeavor we must enlist all possible resources, especially moral, philosophical, and spiritual, and search for inspiration and insight in the cultural memory of human civilization across the world. It is therefore of great significance that the authors of the collected papers explore the "normative side of nature" (Robert Elliot) in the perspective of the great religions, the tradition of ancient China, and latest philosophical theories. They offer a fascinating vision of the challenges ahead and the means to successfully confront them.

Gerhold K. Becker
Editor, Studies in Applied Ethics

# ACKNOWLEDGMENTS

The chapters in this volume grew out of a conference that addressed the environmental issue under the title, "Environmental Ethics: An Inter-religious Dialogue," held at the Centre for Applied Ethics, Hong Kong Baptist University, 9-10 June, 2005. I wish gratefully to acknowledge the financial support provided by the University, the Centre and Wen Hsin Scholarship. Following the conference, the papers were reviewed to produce this volume directed to reflect on environmental ethics from different cultural perspectives. Thus, the volume is entitled "Environmental Ethics: Intercultural Perspectives." I want to express my deep gratitude to Peter Luk Yiu-wang for his invaluable assistance in the various stages of the editorial process. Last but not least, I also want to thank Gerhold Becker, Special Series Editor of VIBS, for including this volume in the series and his unceasing support for the production of this volume.

# ENVIRONMENTAL ETHICS: INTRODUCTION

King-Tak Ip

On 17 January 2007, the Chicago-based Bulletin of the Atomic Scientists (BAS) adjusted the Doomsday Clock from seven to five minutes before midnight, the time for the annihilation of human habitats on earth. The decision was made by the BAS Board of Directors in consultation with the Bulletin's Board of Sponsors, which includes eighteen Nobel Laureates.

The Doomsday Clock was introduced in 1947 to measure the danger of a nuclear holocaust. However, the minute hand of the Clock was moved this time not solely because of the threat of nuclear weapons. The keepers of the Clock expanded its purview, for the first time in sixty years, to include the threat of global warming, the genetic engineering of diseases, and other "threats to global survival." By moving the Clock so much closer to the midnight, the BAS Board of Directors wanted to bring our attention to two major sources of catastrophe that might cause an end to human civilization. These two sources are the perils of nuclear weapons and the destruction of human habitats as a result of climate change.

As the decision indicates the BAS accepts that the threat posed by climate change is as serious as that posed by nuclear weapons. Although the effects of climate change may not be as dramatic as those of nuclear explosion, climate change could bring irreversible harm to the habitats upon which humankind depends for survival. Because climate change brings about increasing environmental deterioration, such damage caused by human behavior has become a global moral issue.

The chapters in this volume grew out of a conference held at the Centre for Applied Ethics, Hong Kong Baptist University, 9-10 June 2005, that addressed the environmental issue under the title, "Environmental Ethics: An Interreligious Dialogue." Following the conference, the papers were reviewed to produce this volume directed to reflect on environmental ethics from different cultural perspectives. Thus, the volume is entitled "Environmental Ethics: Intercultural Perspectives."

How could nature be normative? Robert Elliot attempts to deal with this question from the perspectives of consequentialism, deontological ethics, and virtue ethics. He realizes that the maximizing, improving, and maintaining consequentialisms view the relationships between value and obligation in different ways, but they are common in taking human obligations as nothing more than a matter of calculating values and plugging them either into a maximizing, or an improving, or a maintaining function.

Deontological theories focus on the promotion of intrinsic value and offer duty that do not reduce to function over value. Deontological theories sustain a distinctively environmental ethics with its scope extending beyond the interests and concerns of humankind. Yet these theories run into problems of justifying duty toward those entities that lack consciousness or desire; weighing up, balancing, adjudicating, and ranking conflicting rights; and rendering impermissible actions that seem obligatory in certain situations.

As for virtue ethics, Elliot points out that disputes may arise about what should be put in lists of virtues and vices and how one could decide which is a virtuous or a vicious action. Moreover, it is notorious that virtues carried to excess could change to vices. Nevertheless, he believes that frugality and generosity are the most relevant accepted virtues in the context of sustainability.

He concludes that each perspective leaves room for nature to make direct demands on us for some actions and behaviors. Yet he is less certain about virtue ethics than the other theories.

Leaving room for nature is an imperative in environmental ethics but the human perspective is ineliminable, Gerhold Becker argues. He points out that, in rejecting anthropocentrism, people think that it would neither be reasonable nor justifiable to understand and value nature from the human perspective. However, such a position fails to credit the proper and unique role of humankind in nature.

Becker indicates that respect for nature is the categorical imperative of an ethic which rejects anthropocentrism and seeks to establish biocentrism. Tracing the major stages of the development of the ethics of respect, he concludes that it shares the Kantian basis of human nature as well as similar conceptual problems. He finds it both difficult to claim that all living organisms are ends in themselves and to extend the concept of self to nature.

Becker also points out that it would not be easy to define and justify the value boundaries of nature. It would be meaningless to extend the boundary of nature beyond the planet we inhabit because that would defeat the human efforts to save and sustain the common habitat for both humans and animals. It would be equally meaningless if we discard the concept of nature altogether for the ethics of respect of nature would lose its object if the naturalness of nature no longer exists.

Becker indicates that the prime focus of the ethics of respect for nature is the metaphysical concept of nature. As the Christian theology of creation cannot serve as the basis to argue for the intrinsic value of nature, secular ethics needs to revise its arguments in favor of the metaphysical quality of nature *per se*. However, if the value of nature were so independent that it goes beyond human estimation, then humans would have no way of knowing about or to appreciate it.

Moreover, an ethics of respect should aim to give moral status to nature by making factual claims, such as animals are "conscious creatures" and they are "experiencing subjects of a life." Yet such claims originate from specific human experiences and are difficult to substantiate in animals.

Becker thinks that the concept of nature is a constructed, cultural product. What he attempts to do is to rehabilitate teleological thinking. For him, nature is art, the skillful production of things according to a plan. Nature's goal is in itself, self-preservation and nothing else. Hence, from the perspective of teleology, as art is the second realm of human activity and humankind is part of nature, human beings should hold a sympathetic interpretation and should involve in the self-preservation process of nature.

However, nature could also be viewed by some as a purposeful system and self-conscious being serving as the basis for teleology in all life forms. Such a view elevates and transforms nature into a goddess. Nature has taken over the place reserved for the Christian God. However, could this goddess set the line separating the permissible from the impermissible? Could this be a god whose fate is at the disposal of humankind?

Thus, the ethics of respect for nature cannot be based on absolute values that would define limits to human activity.

While Becker argues for an indispensable role of humankind in doing environmental ethics, Ingmar Persson rejects anthropocentrism as speciesism. He is sympathetic to conativism; something has moral status for being in possession of desires. He accepts that, while the desire to stay alive is stronger than desires for most other things; some animals have stronger and more diversified desires than other animals. Therefore, conativism acknowledges that it is better to kill non-human animals than normal human beings. Nevertheless, this does not imply that conativism agrees on a higher moral status of human beings by virtue of their belonging to the species *homo sapiens*.

Persson believes that human concern for the environment is aesthetic, not moral. For him, environmental ethics is anomalous for, unlike other branches of ethics, environmental ethics does center itself on individuals with moral status. One can act wrongly against beings capable of desires without harming the environment. Since the subject matter of morality is the treatment of entities with moral status, environmental "ethics" is not a branch of ethics.

For Persson, human concern for the environment can tell us about the evolution of the earth and of life on it, and that nature is beautiful, magnificent, and awesome. In order to promote the end of making the world more beautiful, certain individuals could be harmed so as to preserve their species or other species. Such an aesthetic concern is uniquely human for humankind is the beholder of beauty. Thus, what Persson argues for is a consequentialist and aesthetically anthropocentric conception of environmental concern, which he does not regard as a genuine branch of ethics.

While philosophical ethicists attempt to give systematic and commonly agreed methods to settle the environmental questions, Y. S. Lo believes that environmental ethics is empirical and cannot be answered by philosophy alone. Lo explores an environmental empirical ethics. Empirical ethics emerged only recently as a field of interdisciplinary study that takes scientific methodology seriously and attempts to deal with abstract moral questions by using concrete empirical methods. She wants to provide a framework which can devise systematic empirical methods so as to determine what a genuine value is, whether it is universal or relative, and how competing values are to be prioritized.

For Lo, many moral questions are ultimately reduced to empirical questions about human psychological capacities, tendencies, and habits. Her project is both humanist and dispositionalist. Her humanist working hypothesis is that the purpose of ethics is to answer how human beings can co-exist in sustainable, happy, and flourishing ways and to assist them doing so. Her dispositionalist working hypothesis is that core moral concepts can be analyzed and defined in terms of people's dispositions to give affective responses under a set of favorable conditions. She tries to integrate her two hypotheses with the naturalistic and reductive approach of empirical ethics. She thinks that the active application of her humanist dispositionalist theory of value to people in social, political, and moral conflicts will bring modern liberal democracy to its optimal forms of accommodating differences and retaining standards.

In applying her theory to environmental ethics, she points out that if people want to make any environmentally-ethical practice durable, they need to develop both a non-egocentric and a non-anthropocentric sense for virtue in following environmental-friendly conventions and a corresponding sense for vice in violating them. These environmental virtues are dispositions to be internalized to experience different moral sentiments, under certain favorable conditions, toward different objects and behaviors. Non-anthropocentric environmental philosophers need to promote a realistic understanding of human evaluative psychology and try to excite the relevant moral sentiments in others so they can sympathetically understand their philosophical arguments. However, Lo warns, this can happen only if people possess already certain dispositions to appreciate, respect, and care for the environment and to convey a sense of their moral significance. The question about the existence of such original ingredients of the human psychology is an empirical one, which cannot exclusively be answered philosophically.

Apart from chapters written from philosophical and empirical perspectives, chapters from different religious perspectives have been included.

Speaking from the Christian tradition, Michael S. Northcott points out that a new discipline called biogeochemistry is needed to study the complex interactions between atmosphere, land, ocean, and species. The biogeochemists' description of climate change is in accordance with the Biblical story. The Biblical story is a moral and spiritual account of environmental deterioration. However such moral and spiritual chains of causation are absent in the modern language of risk. Modern narratives prefer to use the discourse of risk instead of the discourse of perils and dangers. Yet, this does not mean that nature and pollution are somewhat depoliticized. As risks imposed upon them by the nation state and its corporate partners are largely involuntary, people ironically rely on the nation state to defend them against the risks emerging in the late modern era of "biopolitics." This new politics of risk interacts with genuine environmental threats, which are linked to the decisions and policies uncontrolled by individuals. Due to these imminent threats in late modernity, the political processes are dominated by life politics, and collective fear and anxiety become its cultural character.

Nevertheless, Northcott insists that fear should not be accepted as the basis of a Christian ecological ethic. For him, reverence for God and God's creation should be the foundation. The root of the ecological problem is the modern substitution of human reverence for God and divine providence with a reliance on the sovereignty of reason and the power of science.

With regard to the moral problems of future generations, Northcott argues that it is inadequate to use utilitarian calculation to defend the interests of future generations, since modern moral deliberation assumes that the future is empty. In his view, only Christian ethics can offer an adequate account of the future. Although Christian ethics is intrinsically connected with memory and tradition, by remembering the life, death, resurrection, and ascension of Jesus Christ, Christians believe that the future forms of the created world are already revealed by God. Remembering these stories has become the basis of Christian moral deliberation and the spring of hope and moral courage.

Northcott believes that, in the Christian eschatological perspective, the goal of infinite consumption of energy, which directs modern civilization, is the primary cause of the moral peril. This consumer oriented civilization fosters the disordered human desires and the human illusion that people are free to choose not the gifts of God's earth, but their destinies as individuals and as a species. The risks that human beings face are not only climatological or ecological but also moral and spiritual. The scientific narrative of climate change fails to describe the moral and spiritual perils at stake. What is needed is truthful speech, with speakers ready to bear the burden of wrongs, to relent, and to lay down their life for their friends. Only when the story of climate change is truthfully narrated, can people be expected to response in morally appropriate ways.

From an Islamic perspective, Anis Ahmad believes that the root of Islamic ethics is the revelation, or *wahi*, of Allah. According to the Qur'an, *wahi* is an objective truth with universal character and applicability. For Ahmad, the ethical teaching of the Qur'an and the Prophetic *hadith* and *sunnah* provide general principles, specific instructions, commands, and directives in dealing with global ecological issues.

According to the Qur'an, ecological balance, inter-relatedness, and interdependence as manifested in the principles of coherence and unity (*tawhid*) are taken as signs of willful acceptance of Allah's ultimate authority.

In the Islamic world view, people's lifestyle relates directly to the environmental crisis people face today. Allah is the Creator who sets the ecology in balance, sustains, and promotes life. The Qur'an teaches people to live with a thankful attitude and to live a balanced, moderate, and simple life, though they have abundant resources. The Prophet also requires his followers to protect not only the environment but also animal life.

In the Islamic tradition, *halah* (permissible) and *haram* (ethically prohibited and bad) are the two ethical categories that serve as motivating forces for good and bad behavior. By inferring from these two categories, human beings (1) are encouraged to think critically, through a rational process, the meaning in creation and the relationship between nature and the Creator, Sustainer, and Protector, (2) are prohibited from doing anything harmful to the public good, and (3) are not allowed to do anything that would cause harm or danger to human life.

Ahmad believes that environmental balance and purity lead to social, economic, political, and cultural equilibrium and balance. Concern for the environment ultimately nurtures an ethically-motivated behavior, a moral force, in individuals and society.

In attempting to construct a global ethical force, Ahmad provides a road map for the betterment of ecological and environmental development. What he offers is a sevenfold strategy. The seven universal ethical principles he proposes are: (1) developing a holistic approach to life, (2) embracing equitable, just, and fair attitude and behavior toward human beings and the environment, (3) promoting life, (4) supporting inter-faith harmony, (5) keeping rational liberty or liberty of the will as the basis of ethical judgment, (6) preserving dignity and honor, and (7) providing economic and personal security to members of society.

From the perspective of the Pali Canon of early Buddhism, Pragati Sahni examines some challenges to the claim that Buddhist literary sources support an environmental ethics and attempts to establish more progressive forms of Buddhist environmental ethics.

Early Buddhism supports environmental protection. All harming and killing are offensive to the Buddhist practice of non-violence and the killing of the environment is the worst. Apart from protecting animals from being sacrificed, injury to seeds and crop is not allowed. Moreover, Buddha and his disciples enjoyed the beauties of nature. At a more fundamental level, Buddhism advocates the cleansing of the human mind. Humans need to remove from their mind greed, hatred, and stupidity, the three "poisons" that the Buddha spoke of. Without the mind being cleansed, the environment would not be purified.

Sahni also notes, however, many problems with regard to the environmental ideas of early Buddhism. Even though Buddhism envisioned a harmonious relationship between nature and humankind, Buddhist principles do not include any modern ecological ideas, and Buddhist soteriology implies a world-denying attitude of life. In addition, while Buddhist texts are often quoted with a view to supporting the protection of nonhuman species and environment, closer examination shows that most of the texts can be interpreted in ways that are indifferent to or even contradict the spirit of environmental ethics. Therefore, scholars attempt to use the hermit strand and other Buddhist doctrines such as the notion of self, or of dependent co-arising as alternative resources for a Buddhist environmental ethics.

Summarizing these supportive and problematic factors within the early Buddhist texts, Sahni explores two possibilities, which are inspired by deep ecology and environmental virtue ethics, in establishing an early Buddhist environmental ethics.

For Sahni, deep ecology's concern with eliminating harmful ideology and reframing human thought process finds resonance in early Buddhism. However, early Buddhism does not accept the notion of "interdependence" without boundaries as deep ecologists do. While it opposes human superiority, it does not reject anthropocentrism as fervently as deep ecology dose. Hence, early Buddhism shares similarities with, yet also poses challenges to deep ecology.

Sahni points to an overwhelming concern about character building in early Buddhism. The doctrine of *kamma*, which teaches the inevitable cycle between an action and its consequences, is closely related to the character of the moral agent. The Pali Canon emphasizes on the importance of mediation, wisdom, and morality in the removal of greed, hatred, and delusion so as to achieve the ultimate goal of liberation or *nibbāna*. Therefore, both *kamma* and the pursuit of liberation require one to be trained as a virtuous person. Although the goal of Buddhism is not the improvement of the environment, all aspects of a person's life, including the treatment of the environment, will inevitably be affected by the internalization of Buddhist virtues.

Though adopting a virtue approach looks promising, it has its problems. The first problem is the obscurity in defining what can or cannot be taken as environmental virtue in early Buddhism. The second is the lack of guiding principles in resolving conflicts between contrasting virtues. While accepting the challenges posed, Sahni points out that adopting the virtue-ethics approach allows to respond to some of the problems faced by Buddhist environmental theory. Nevertheless, such a theory can never be treated as a stand alone theory for it is inherently merged with the Buddhist goal of liberation.

Sahni admits that many problems remain after exploring the possibilities inspired by deep ecology and environmental virtue ethics. Nonetheless, a feasible Buddhist environmental ethics needs to be set up through a careful scrutiny of such possibilities emerged.

Jonathan Chan argues that Daoism advocates an environmental ethics with a sharp focus on sustainability. He shows that a virtue ethics needs to place sufficient emphasis on the normativity of environmental virtue. He argues that Daoism can provide powerful support for the normativity of environmental virtue.

For Chan, naturalness, which regulates all other Daoist virtues, is the first virtue and of utmost importance. Nature manifests *Dao*. Human beings have to learn from nature. The Daoist human-nature relationship requires humans to pay respect to nature and to cultivate such a disposition in them. Daoists believe that the one who exemplifies Daoist virtues is the sage. The sage is an environmentally virtuous person who champions the naturalness of all things. In order to do so, one needs to take a non-interruptive approach as required by the Daoist virtue of "*wuwei*" (or non-action).

The virtue of *wuwei* (non-action) shows a strong version of sustainability, which does not allow exploitative policies or practices which interrupt nature extensively and lead to the deprivation of the environment.

While Chan argues for a Daoist view of environmental virtue ethics and a strong Daoist notion of sustainability, Livia Kohn examines how the Daoists realize their understanding of the human-nature relationship. Kohn points out that, according to Daoism, the universe is good and functions in perfect harmony. The universe is created from the pure and formless *Dao* and manifests itself in an energetic life force known as *qi*, the cosmic force that causes things to live, grow, develop, and decline. Ideally, *qi* works to make up a cosmos of perfect good. When it works with nature, human beings experience a sense of well-being and inner harmony.

For Kohn, *Dao* or "the Way" is the most essential notion in Daoist environmental ethics. Through *wuwei* (non-action) and naturalness one lives with *Dao,* and adjusting one's way of life according to *qi*, the basic stuff of nature, is the best way to reach *wuwei*. Thus, Daoists must live a life in which perfect balance is maintained between *Dao* and *qi* and in human life and the natural world.

In order to work out such a way of life, Daoists must practice *Fengshui* so as to put themselves in the most auspicious space. They should also live in line with the practices as required by the traditional Chinese calendar for, like space, time is a form of *qi*-manifestation and has to be received and set off accordingly. Apart from space and time management, they have to follow the Daoist communal code as stated in the Daoist classics. These moral rules reflect a central concern with *qi*, which should never be lost but always be guarded, preserved, strengthened, and cultivated in nature, state, and family.

According to Kohn, research in biology, physiology, and physics can give Western science the language to explain Daoist concepts and to make them widely accessible to the public. Daoist wisdom may draw public attention to the environmental problems and start the long healing process.

Environmental protection is a global issue. What lies before us is a task concerning to life and death of the Earth community. The essays collected here represent a joint effort in dealing with this problem. All contributors to this volume agree that what we urgently need now is global awareness of the environmental crisis we are facing. I hope this volume will serve as a catalyst in stimulating more debates on environmental questions across cultures.

# One

# THE NORMATIVE SIDE OF NATURE

Robert Elliot

## 1. Introduction

How could nature be directly normative, and how could its normativity have bearing on policies of sustainability? One answer is that nature is not directly normative, and that the only justifiable environmental ethic is a human-centered ethic. On this view, decisions about how to treat nature are made solely with reference to human interests. For example, because thc retention of tropical rainforests or wetlands is crucial in maintaining human health and well being, this provides a reason for conserving nature. The normative force of the conservationist policy, though, derives from the claims of human interests on moral agents, not from nature directly. Nature enters the picture only indirectly or hypothetically. Provided human interests are appropriately dependent on the conservation of nature and natural items, the goals of sustainability will converge with the goals of environmentalism. Indirect considerations are potentially quite compelling, and are more motivating for most human beings than direct considerations.

Could nature be normative in a direct way? One suggestion might be that nature has intrinsic values that demand its conservation. It might be argued that natural values exist, such as wildness, which are relevant to moral choice in much the same way as are more familiar, human-centered values, such as well-being, happiness, or health. These intrinsic values might be thought to direct moral agents to advocate and pursue policies of sustainability, assuming that sustainability tends to require the conservation of nature. This first suggestion about nature's normativity is best thought of as a form of consequentialism, where the values relevant to the assessment of consequences go beyond those associated with human beings.

Another suggestion might take the view that nature has ends that have a direct normative force that constrain the choices of moral agents. The idea is that nature's ends, the ends of natural entities, are deserving of respect in something akin to the way in which human being ends are said by some to be deserving of respect. This general suggestion about the normative side of nature could also be couched in terms of the language of moral rights. The rights that are attributed to nature would derive from nature's claim to be respected from its constraining normative side. Those taking this view would likely find policies of sustainability appealing, because such policies result in more extensive respect for nature and natural entities, or taking nature's rights seriously to a greater degree. This second suggestion is usefully thought of as a deontological view of nature's normativity.

A third possibility is to connect nature's normativity to virtue ethics and the idea of virtuous action. Virtue ethics assumes that there are virtues and vices that provide us with a guide to action. We are enjoined to act virtuously and to shun vice. Plausible examples of virtues are bravery, compassion, generosity, humility, and frugality. Plausible examples of vices are greed, self-indulgence, cruelty, dishonesty, cowardice, and arrogance. However, if virtue ethics is to provide the basis of nature's normativity, then it must be that nature itself in some way compels the virtuous actions. It cannot just be that being virtuous regarding human beings requires us to adopt policies of sustainability.

Each of these suggestions may be restricted in different ways. Each might attribute normative consideration to nonhuman animals, each might attribute normative consideration to all living things, or each might attribute consideration to all natural items whether living or non-living. Each of these suggestions puts nature, or parts of nature, in the driver's seat. The normative side of nature derives from the attribution of moral properties, or moral force of some kind, to nature itself. In developing this general idea we might take the view that the normative side of nature is completely and deeply independent of human desires, preferences, or evaluations. This would be the view that nature objectively has a normative side. Or we might take the view that ultimately the normative side of nature depends on human desires, preferences, or evaluations. This would be a subjectivist view of nature's normativity. On both of these views nature is directly normative, although they take different positions on the basis of that normativity. We can adopt the normative view that nature has characteristics that, prima facie, require us to act in some ways, independently of the impacts on human beings, but remain uncommitted on meta-ethical issues as to the fundamental nature of moral properties, values and the like.[1]

## 2. The Consequentialist View of Nature's Normativity

Consequentialism defines the obligatory as a function over intrinsic value. As considerations to do with intrinsic value, the value that something has in itself or for its sake, are taken to exhaust the normative content of consequentialist principles of obligation. These considerations define what, in consequentialist terms, is permissible, obligatory, and impermissible. Other moral concepts, such as duty, will be similarly defined on the consequentialist view via the concepts of permissible, obligatory, and impermissible.

Consequentialism enjoins examination of the consequences of actions in determining whether they ought to be performed, ought not to be performed, are right, wrong, obligatory, permissible, and so on. It requires examining the intrinsic values and disvalues attaching to those consequences and to the actions themselves. The most common form of consequentialism directs us to maximize intrinsic value. Another form worth discussing directs us to increase intrinsic value, and another directs us to maintain intrinsic value. These variants express obligation generally as a non-maximizing function over value: inserting the relevant values specifies particular obligations.

Consider, maximizing consequentialism. It tells us that an action is obligatory if, compared to the other actions that it is open to an agent to perform, it maximizes the expected quantity of intrinsic value, and further, that only those actions that maximize expected intrinsic value are permissible. The obligatory and the permissible coincide. Utilitarianism is perhaps the best known maximizing consequentialism, recognizing only pleasure or happiness as intrinsic values. It obliges us to maximize expected pleasure or happiness, other actions being impermissible.[2] Utilitarianism is only one possible kind of maximizing consequentialism. Other versions are identified by the different intrinsic values they specify. The practical outcome of applying these principles of obligation is determined by these intrinsic values. According to some theorists, there are distinctively natural intrinsic values that are exemplified by nature in its relatively unspoiled states.[3] According to maximizing consequentialisms, if failure to preserve wild nature delivers a less than maximal increase in intrinsic value, then there is an obligation to preserve it. Similarly, if restoring a natural area that has been degraded maximizes expected intrinsic value, then there is an obligation to restore it. With all serious variants of consequentialism, the support for environmentalist policies is strongest where nature itself is taken to have intrinsic value. The support will be exceptionally strong where natural values are the only intrinsic values recognized, although a normative environmental ethic that counts only natural values as intrinsic values would be immensely controversial. Support would also be quite strong where, even though other values, including human happiness or well-being, are recognized, natural values are taken to be the most important values. This will be especially so given the reasonable assumption that long-term human happiness and well-being require us to conserve nature.

A further aspect of consequentialisms should be especially stressed; namely that they permit trade-offs between quantities of the same intrinsic value and between different intrinsic values. For example, if pleasure is the only intrinsic value, then a maximizing consequentialist would say that it is obligatory to reduce one person's pleasure, or even to inflict pain, in order to maximize pleasure overall. A consequentialist who accepts that there is a plurality of intrinsic values, including natural values, is faced with a quite difficult task: having to make comparative judgments not only about different quantities of the same value, but also about different quantities of different values. The consequentialist who thinks both pleasure and the acquisition of knowledge are distinct values has the problem of deciding just how much pleasure should be sacrificed in order to advance some particular area of knowledge, and vice versa.

Because of the possibility of trade-offs, it is useful to say that if some natural item has intrinsic value, it is prima facie obligatory not to destroy it, instead of obligatory to do so, all things considered. Noting that something is of intrinsic value puts us on a warning not to destroy or degrade it, unless it can be shown to be the case that such an action maximizes expected value. It is always necessary to ensure that there are no alternative actions that might be performed that would increase value to a greater extent. Consequentialism takes a global and long term perspective in delivering its normative assessments. Establishing that some action is permissible, all things considered, requires a serious attempt to assess its impact far from its geographical location and into the further future. This is especially pertinent where natural ecosystems and processes, upon which all life and well-being ultimately depend, are involved.

When relevant consequences, values, and alternative courses of action are assessed and compared, we might discover that the loss of intrinsic value through environmental despoliation could be, in principle, compensated for by increases of intrinsic values that are human-centered. For instance, environmentalist policies are often countered with the claim that the development of some natural area will result in substantially increased benefits for human beings, such as employment opportunities or increased material wealth. The implication is that environmentalists have their value priorities wrong, or that they are illegitimately discounting the substantial benefits for human beings that flow from environmental destruction. The possibility of this style of argument is one reason that many environmental ethicists have been cool toward consequentialism, looking elsewhere for a framework for their normative beliefs.[4]

Consquentialism, provided it recognizes natural values, is in a strong position to defeat this style of argument. This is because such arguments typically exaggerate benefits for human beings, underestimate deleterious environmental impacts, ignore alternative means of benefiting human beings, do not investigate alternative social and economic arrangements, underestimate the costs of environmental despoliation and degradation to present and future human beings and nonhuman beings, and fail to interrogate the connections between quality of life and material wealth. If the consequentialism in question recognized only natural value, although that would constitute an extremely controversial value theory, or gave natural values special significance or intensity within a plurality of intrinsic values, then the concern that consequentialism is an inadequate normative environmental ethic should be allayed.

A feature of maximizing consequentialism, noted earlier, is that it does not allow any deep distinction between the obligatory and the permissible. This is sometimes thought to be a weakness of maximizing consequentialism in that, allegedly, the theory asks more of people than a normative ethic could reasonably ask. Some variants of consequentialism do allow that actions that are permissible need not be obligatory. Assume, for example that there are natural values, and consider improving consequentialism, which says that it is obligatory to act so as to increase, although not necessarily maximize, intrinsic value. Such a view is not all that odd. It reflects the plausible maxim that we should leave the world better than we found it.

This variant of consequentialism is less onerous than the maximizing variant, requiring less of agents than a total all out effort to maximize value. Environmentalists, who were concerned by the trade-offs implicit in maximizing consequentialism will be concerned here too. There are cases where some environmental despoliation may lead to an increase in value, but in which some alternative action, not involving environmental destruction, would increase value to a greater extent. Improving consequentialism appears to leave open the possibility that the first action is permissible and the second not obligatory. There are three reasons for thinking that the implications of adopting an improving consequentialism are not, from an environmentalist perspective, quite so worrying.

First, improving consequentialism does not require us to do the best we can. It requires us only to improve matters to some degree. If, a person is willing to make a degree of effort to improve things, we might reasonably require that person put effort to produce the best result possible. Improving consequentialism might reasonably be taken to have an efficiency requirement built into it. The upshot would be an obligation not merely to ensure that an action improves value, but also to ensure that no alternative action involving the same degree of effort improves value more. An improving consequentialism that recognized intrinsic natural values, and took them to be intense, would strongly favor environmentalist policies. Second, improving consequentialism also takes a global and long term perspective, requiring agents to take pains to ensure that their actions are improving from that perspective. Relatedly, improving consequentialism must engage in an honest appraisal of the consequences of those actions embarked on in order to improve value. Third, in improving consequentialism, there is considerable looseness involved, because improvements may be quite large or quite small. Serious advocates of improving consequentialism might think their position is trivialized if acceptable improvements need only minimally increase value. Instead, they might insist, still with a degree of looseness, which significant improvements are required.

Another, still less onerous, variant of consequentialism makes it obligatory to act so as to maintain, although not necessarily increase let alone maximize, intrinsic value. Maintaining consequentialism might be thought to provide little support for environmentalist policies, but initial impressions might be misleading here, especially where natural values are taken to be intensely significant. It is not difficult to imagine the maintaining consequentialist trying to maintain intrinsic natural value in the face of extensive and recurring acts of environmental destruction. Actions, including those which impact directly on wild nature, will, all too frequently, depress intrinsic value. Acting to promote environmentalist policies will be a clear and obvious way to fulfill the requirement to maintain intrinsic value. Ironically, an onerousness objection might be pressed even against maintaining consequentialism, since in a world in which the loss of intrinsic natural value proceeds apace, the requirements of even maintaining consequentialism may be exceptionally demanding.

The three variants of consequentialism considered involve differing relationships between value and obligation. They have in common, the view that figuring out our obligations is nothing more than a matter of calculating values and plugging them into some function, such as a maximizing one, an improving one, or a maintaining one. Each provides a framework for a normative environmental ethic, and their adequacy in this regard is crucially dependent on the specific values that are recognized, and the comparative primacy given to natural values. Nor should we lose sight of the range of possible values, including those that reflect the interests of human beings and other sentient creatures. The assault on the natural environment massively threatens these interests, and a consequentialism that recognizes them would compel serious environmentalist policies.

## 3. The Deontological View of Nature's Normativity

Let us turn to another major ethical tradition that is constituted by deontological normative ethics. Deontological ethics are often characterized as ethics of principle instead of ethics focused on promoting intrinsic value. Unlike consequentialist theories, they offer principles of obligation or duty that do not reduce to functions over value, allowing the judgment that actions are obligatory for reasons in addition to the value of their consequences. Deontological theories claim that some kinds of action are obligatory, permissible, impermissible, and so on, in virtue of specific, nonconsequential properties of that action. They do not necessarily exclude such axiological or value assessments, and complete deontological assessments may require some prior axiological assessments.

It might be claimed that since some natural object has intrinsic value, an obligation exists not to destroy it. The property of being destructive of a thing with intrinsic value would, according to this ethic, be a wrong making property; the relevant maxim or principle being "do not destroy things which have intrinsic value." This maxim has about it the flavor of a consequentialist principle. The normative assessment is not carried out by calculating the loss of intrinsic value associated with the destruction of the object and figuring it into some principle of obligation, such as those variants of consequentialism earlier considered, which is as a function over value. Instead, the wrongness of the act can be established without having to look beyond the fact that it involves destroying something of intrinsic value. There is no suggestion that people need look to the consequences of such acts, or that people ought to act in accordance with some function of the intrinsic value of the consequences of the act, and that of its alternatives.

There is no suggestion that it is permissible to destroy something of lesser value in order to protect or create something of greater value, which is one reason some environmentalists have felt less unease about deontology than consequentialism. The difference is akin to the difference between a principle that enjoins us to minimize pain, which is consequentialist, and a principle that forbids us to cause pain, which is not consequentialist. It may be impermissible to act in ways that maximize, improve, or even maintain intrinsic value; for instance, in situations where the only means of doing one of these things involves the destruction of something of intrinsic value contrary to the prohibition on destroying such. A deontological structure exists that would sustain a distinctively environmental normative ethic, the scope of which extends beyond human interests and concerns. For destroying or degrading the natural environment could be wrong, because it is an act of destroying things that possess natural intrinsic value. The wrongness does not result from the reduction of value as such: the wrongness results from an independent non-consequentialist principle.

There are other ways of fitting a distinctively environmental ethic into a deontological structure. Theories of rights are often presented as deontological theories, because they imply the proscription and prescription of acts independently of the consequences of those acts. As someone's right to life might be said to result in an absolute proscription on taking that person's life, except perhaps in self-defense or in a judicial context, irrespective of the consequences. Because value is increased as a consequence it is not, it is often claimed, an acceptable justification for violating the right. Much environmental ethics might be cast in terms of rights. Most obviously, it makes sense to invoke the rights of nonhuman animals in objecting to the destruction of natural habitat. Some have wanted to extend the concept of rights beyond the set of sentient creatures suggesting that, in addition, plants have rights that species have rights, or that ecosystems have rights. This proliferation of rights generates problems. First, there is the issue of whether the extension of rights in these ways is conceptually sound.[5] Does it make sense to attribute rights to entities that do not even have desires and that are not even conscious? Do we even want to suggest that non-living natural items, such as rocks or glaciers or rivers, could have rights?

Equally crucial is the practical problem of how to process and adjudicate the barrage of rights claims that would be generated by such deontological ethics. The problem would be ameliorated if we could be sure that the rights in question would not conflict, but that is not at all clear, even where we are focusing only on the rights of human beings. In the context of extended rights theories, conflicting rights seem inevitable with attendant problems of weighing up, balancing, and adjudicating countless, apparently competing, rights claims. The problem appears more acute for a deontological theory than a consequentialist theory, because the first eschews trade-offs based on consequences. How do we respect the rights of every organism? Is there a hierarchy of rights? Is there a hierarchy of rights-bearing individuals, such that the rights of humans have priority over the rights of sentient nonhumans, which have priority over the rights of other living things? The answer, even in theory, is not clear, and the ethic that suggests the principle might therefore be thought vacuous. The prospect of vacuousness is brought out if we consider the claim, often associated with deep ecology that every living thing has an equal right to flourish. Life on earth is such that particular organisms can flourish only if others do not. Taking the right appears to leave no room for action.

Some are tempted to say that the problem just sketched is the general one that affects ethical extensionism as the method for generating an environmental ethic, namely that things go awry when we focus on individual entities at too fine-grained a level. Such theorists might suggest that we should be focusing on macro-entities such as whole ecosystems or the biosphere as the pertinent rights bearers. This move might stem the proliferation of rights but it still leaves the problem of how to make sense of the claim that entities that lack consciousness or desires could have rights. No parallel problem in the suggestion exists that they have intrinsic value, and so no problem in a deontological theory that prohibits the destruction of what has intrinsic value. It is odd to suggest that they have rights in the sense that humans and sentient nonhumans have rights. For one thing, unless an entity is conscious, there appears no content to the suggestion that from its point of view things are going well or badly. The point of rights theories appear to be to create a set of entitlements on the part of individuals that allow things to go well from an individual's point of view.

There would still be a residual ranking problem in working out the respective priorities of the rights of sentient creatures, ecosystems, and the biosphere. A simple solution would be to give absolute priority to biospheric rights. This solution would be unpalatable to many, because it would demote human rights to little more than an afterthought, making human interests subservient to those of the biosphere. Perhaps, this is an idea that we could get used to if we are convinced of the intense ethical significance of the natural environment. While there are limits to the capability of a deontological theory based on rights to support the moral sentiments expressed by many environmentalists, such a theory can accommodate many. Acts of environmental destruction and degradation will be wrong for human-centered and animal-centered reasons that a deontologist would likely find compelling. For example, such damage would wrongfully injure and kill nonhumans, and wrongfully impose costs and burdens on humans, including future humans. The attendant ethical concerns can be powerfully and coherently expressed in the language of rights.

A deontological theory exists, Kantianism, deriving from the views of the Immanuel Kant, which is similar in structure to the rights-based theories and which deserves some comment. The central tenet of Kantianism is that persons are an end in themselves, having a capability for rational autonomy requiring respect as a person. The idea of respect for persons might be thought to be the basis of theories of rights that articulate and elaborate the idea of respect for persons. At first sight, Kantianism, emphasizing as it does respect for persons, might not seem to provide an amenable structure for any thing much more than a human-centered environmental ethic. One prominent theorist, the American philosopher Paul Taylor, has elaborated a normative environmental ethic with a Kantian flavor. Taylor asks us to scc all living things as autonomous in that, they have biologically based goals that are definitive of the kinds of organisms they are and that define for them what counts as flourishing.[5] He suggests that, just as traditional Kantianism enjoins us to respect the rational autonomy of persons, so too a naturalized Kantianism enjoins us to respect the natural autonomy of all living things. The force of Taylor's position derives from whatever success he might have in convincing us that a useful analogy exists between rational autonomy and natural autonomy, and our views about the significance of rational autonomy. Taylor does not seem to want natural autonomy to swallow up rational autonomy, seeking instead to maintain a moral distinction, with hierarchical implications, between persons and other living things.

The conceptual and proliferation problems that affected rights-based deontological theories are present in Taylor's theory. The analogy between rational autonomy and natural autonomy might well founder on the fact that so much of the second involves no consciousness of preferences or desires. Although we might well see the point of allowing that non-sentient living things have a kind of autonomy, we might think the conceptual distance between the autonomy of an orchid and that of a primate is too great to sustain the mooted ethical extension. The theory runs into to problems of ranking claims based on natural autonomy. How, for instance, do we adjudicate situations in which human welfare is promoted, or rational autonomy is protected at the cost of destroying entities such as plants and microbes that have natural autonomy? One response to these problems is to try to render Taylor's insights in a non-Kantian form. We might accept that something ethically significant about natural autonomy exists, but suggest that its significance is best articulated through the concept of intrinsic value. We can say that natural autonomy is a basis of intrinsic value, and either plug that into a consequentialist framework, or into a non-Kantian, non-rights based deontological framework.

A final problem exists that should be sketched. In an uncompromising form, deontological ethics might render impermissible actions that do not seem impermissible and that may even be obligatory. For instance, the degradation of some small area of the natural environment in order to create a firebreak may be necessary to ensure the protection of an extensive area. If what we value is wild nature, then it is permissible to make the firebreak even though it involves the destruction of items of value. A strict deontology is likely to deliver normative conclusions that are difficult to accept. One response, not unproblematic, is to suggest a mixed ethic containing both consequentialist and deontological components. If enough of value is at stake, then it may be judged permissible to act in a way that a strict deontology would proscribe. By the same token, the deontological component would act as a brake on consequentialist justifications of environmental degradation.[6]

## 4. The Virtue Ethics View of Nature's Normativity

As indicated above, virtue ethics has to do with mapping out ideals of human excellence, ideals of good or virtuous character and their opposite, and some plausible virtuous and vicious character traits. There will be disputes about how lists of virtues and vices are to be filled out. For example, some might argue that love of nature be added to the list of virtues, but this will not be universally endorsed. There will be disputes as to whether a particular action exemplifies virtuous or vicious action. Even having a list of virtues and vices about which there is reasonable agreement does not guarantee that a person can straightforwardly classify an action as the exemplification of a virtue or a vice. An apparent act of generosity might turn out not to be so, perhaps because it is motivated by a desire to manipulate another, a desire to have others see one as more important, or because the action imposes an utterly insignificant cost on the agent. Notoriously, virtues carried to excess can transform to vices. For example, actions that exemplify courage, a virtue, might be carried to the point where they exemplify recklessness, which is, according to some, a vice.

Virtues collectively constitute ideals of human excellence, and in a way virtue ethics are more down to earth than consequentialisms and rights-based theories. The key terminology they use is more accessible, and there is maybe more give and take in the pronouncements of virtue ethics. For example, there is no outright suggestion that generosity is obligatory. Instead, it is held up as an ideal kind of behavior to which we should aspire and which we should train ourselves to exemplify as second nature, as a matter of habit. Even if virtue ethics may be a little more forgiving than consequentialism or rights-based theories, it is not weak or wishy-washy. Virtue ethics typically endorse kinds of behavior that require large changes to the way that many of us live our lives: changes that we will resist without being able to defend that resistance rationally, retreating, instead, to self-interest.

In the context of sustainability, the most pertinent accepted virtues would seem to be frugality and generosity.[7] The one implies a careful use of resources with a concern to avoid waste and over-consumption, and the other implies a willingness to assist those whose circumstances are less blessed than one's. Let us consider the first of these virtues. It might be seen by some, as reflecting a view about desirable character independently of anything outside the agent themselves. The idea is that frugality is a virtue because being modest in one's use of resources and the like is an ideal of human excellence, irrespective of the quantity of resources available. What drives this view of frugality being a virtue has nothing to do with leaving enough for others or matters beyond the agent themselves. Frugality might also be seen as a direct response to nature. We might see that frugality in our use of resources means that we do not overly disrupt or disturb nature. The idea is that nature's characteristics demand frugality, and that once we understand nature's characteristics, we will see that frugality is required of us. The situation is like the case of kindness, which is arguably a virtue, the exemplification of which is demanded of us by the characteristics of those around us. It is demanded as a response to the natures of others: it is not desirable in some sense that leaves the virtuous agent disconnected from those upon whom their actions impact.

## 5. Conclusion

I have tried to show how sense might be made of the view that nature directly demands some actions and behaviors of us. I believe that each of the three broad views considered leaves scope for this, although I am less certain about the third view discussed than about the other two.

## NOTES

1. Robert Elliot, *Faking Nature: The Ethics of Environmental Restoration* (London: Routledge, 1997).

2. Peter Singer, "Not for Humans Only: The Place of Nonhumans in Environmental Issues," *Ethics and Problems of the 21st Century*, eds. K. E. Goodpaster and K. M. Sayre (Notre Dame, Ind.: University of Notre Dame Press, 1979), pp. 191–206.

3. Elliot, *Faking Nature.* See, Holmes Rolston III, *Environmental Ethics: Duties to and Values in the Natural World* (Philadelphia, Penn: Temple University Press, 1988). And see also, Richard Sylvan and David Bennett, *The Greening of Ethics: From Human Chauvinism to Deep-Green Theory* (Cambridge, UK: Whitehorse Press, 1994).

4. Sylvan and Bennett, *The Greening of Ethics.*

5. Paul Taylor, *Respect for Nature: A Theory of Environmental Ethics* (Princeton, N.J.: Princeton University Press, 1986).

6. Sylvan and Bennett, *The Greening of Ethics.*

7. Thomas E. Hill, Jr, "Ideals of Human Excellence and Preserving the Natural Environment," *Environmental Ethics*, 5 (1983), pp. 211–224.

# Two

# *JE SUIS LE GRAND TOUT*: RESPECT FOR NATURE IN THE AGE OF ENVIRONMENTAL RESPONSIBILITY

Gerhold K. Becker

## 1. The Biocentric Turn in the Ethics of Respect

"Respect for nature" is the categorical imperative of an ethics that emphatically and exclusively revolves around nature. This ethics not only takes the challenge of continued exploitation and destruction of nature and the natural environment to new heights, but also offers a new moral perspective that stands in direct opposition to traditional ethics. It has gained significance for contemporary ecophilosophy that reaches far beyond its first programmatic exposition in Paul Taylor's well-known book by the same title.[1] Respect for nature calls for a specifically environmental ethics, which no longer takes humanity as its point of departure and as its locus of value and worth but nature. It rejects anthropocentrism, denounces claims of human superiority, and seeks to establish biocentrism as the exclusive normative perspective.

Delineating only some sort of ideal type of such environmental ethics of respect, I suggest to find its main characteristics in a theory of intrinsic worth or in other terminology of intrinsic value of nature, most notably bound up with a teleological account of living beings as ends in themselves, but with an apparent tendency to extend the scope of moral concern ever further.

Taylor develops his theory of inherent worth of nature in direct opposition to but in structural symmetry with the traditional ethics of respect of Kantian provenance. As will be recalled, in Immanuel Kant's view, common morality encapsulates an understanding of human beings as unique in that their worth is "beyond price." Such beings have dignity which commands respect (*Achtung*). Respect is not a natural feeling "received by means of influence," but the feeling that corresponds to rational self-constraint, and that is "self-wrought by means of a rational concept."[2] It represents "our fundamental experience of objective value . . . directed at something whose worth we recognize by reason from within ourselves.[3] Respect in this sense is what Stephen Darwall has called recognition respect and distinguished from appraisal respect: "To have recognition respect for someone as a person is to give appropriate weight to the fact that he or she is a person by being willing to constrain one's behavior in ways required by that fact." In contrast, appraisal respect "is an attitude of positive appraisal of a person either judged as a person or as engaged in some more specific pursuit."[4]

The dignity accorded to human beings obligates us to respect them unconditionally not because of any individual merit, but just because of their being human. Their human rational nature grounds autonomy and distinguishes them as ends in themselves. It implies a qualitative and morally relevant difference between human and non-human beings, most importantly non-human animals; it claims that human dignity is not derived from anything external to the very nature of human beings, but is intrinsic to what makes them human.

Historically speaking, the idea of human dignity and respect has integrated Greek, especially Stoic, and Judeo-Christian intuitions. Martha Nussbaum argues that the "idea of universal respect for the dignity of humanity in each and every person, regardless of class, gender, race, and nation an idea that has ever since been at the heart of all distinguished political – thought in the Western tradition – is, in origin, a Stoic idea."[5] On this account, the Stoic foundation of human dignity is reason, the spark of divinity in human beings, that "marks humans out as incomparably higher, and worthy of a boundless respect and self-respect."[6]

Amalgamating Christian conceptions of personhood, the notion of human dignity took again center stage in Renaissance philosophy. When, in his famous speech on the dignity of humanity, Pico della Mirandola has God calling the human person the "maker and molder of yourself" and "a second god," he recalls almost verbatim the Stoic philosopher Epictetus who also imagined God as saying: "We have given you a certain portion of our self, this power of pursuit and avoidance, of desire and aversion."[7] This suggests that above anything else, human dignity is grounded in the moral nature of human beings presupposing a metaphysical basis of freedom.

The Christian tradition puts the difference between humanity, and all other beings in humankind's creation in God's own image, which provides the idea of human dignity with its onto-theological basis. The implications of this idea were only worked out in the thirteen-century reflections on the personhood of Jesus Christ. The dignity of the human person was no longer conceived along the premise of the Aristotelian ontology of substance (that is, a descriptive conception of personhood, distinguished by certain natural characteristics), but for the first time grounded in humanity's moral nature as *ens morale*, that is, in a normative conception of personhood.[8] The human person is essentially a moral being of freedom and as such has dignity which requires respect.

Kant identifies the essential characteristics of humanity with the "capacity to set oneself an end."[9] "Humanity" and "rational nature" become interchangeable terms grounding moral worth and dignity. By setting ends for ourselves, we set values, and this "capacity" to generate values is what makes us unique and confers on us dignity. Since we are the source of all value there is, humanity itself is of unconditional and incomparable value and must always be respected as such. In his *Groundwork of the Metaphysics of Morals*, Kant summarizes this line of thought as follows: "Rational nature is distinguished from others in that it sets an end for itself . . . . It is that which must never be acted against, and which must never be valued merely as a means, but in every volition also as an end. Now this end can never be other than the subject of all possible ends themselves."[10]

The reason for recalling the major stages in the development of the ethics of respect is a structural symmetry to Taylor's ethics, which shares with the Kantian approach its basis in human nature and inherits similar conceptual problems. In detaching the traditional account of moral standing from its exclusive anchorage in human personhood and applying it to all living organisms. Taylor seems to endorse the Kantian ethics of human dignity and respect, but at the same time also to magnify its problems by presupposing a common moral basis for respect in humanity and all other living organisms. In contrast to an utilitarian focus on sentience as the common characteristic of humans and animals, Taylor's account of inherent worth is burdened with the factual claim that all living organisms are ends in themselves, and in this fundamental regard not qualitatively different from each other. They must therefore be accorded the same moral respect, and any competing interests must be decided exclusively on moral principles.

Taylor distinguishes between intrinsic value, which is "relative to and dependent upon" the valuation of appraiser, and intrinsic worth, which is grounded in objective properties of moral subjects, the most important of which is that they "are teleological centers of life." While he explicitly endorses Tom Regan's claim that at least all higher animals must be regarded as "experiencing subjects of a life,"[11] he extends this claim to all living organisms including the plants: "All organisms, whether conscious or not, are teleological centers of life in the sense that each is a unified, coherently ordered system of goal-oriented activities that has a constant tendency to protect and maintain the organism's existence."[12] From this fact, Taylor derives the "prima facie duty to promote or preserve the entity's good as an end in itself."[13] The required moral attitude of respect for nature is comprised of valuational, conative, practical and affective dispositions toward all beings possessing inherent worth, which even extends to "all living things in the natural world." It entails "the principle of species-impartiality" (implying strict neutrality "between predator and prey, parasite and host, the disease-causing and the diseased") that precludes counting "the good of some as having greater value than that of others."[14] Genuine respect for nature obligates us to act only (or to decline to act) "out of consideration and concern for the good of wild living things."[15] Reconfirming the Kantian tradition, respect is therefore "different from the love of nature" and instead of echoing merely an emotional appeal requires the adoption of valid moral norms that can be justified only within what Taylor calls "the biocentric outlook on nature."[16] "Given this world view, the attitude of respect is then seen to be the only suitable, fitting, or appropriate moral attitude to take toward the natural world and its living inhabitants."[17]

One of the most radical versions of biocentric ethics anticipates Taylor's by several decades. As early as 1923, it was persuasively argued (had already been put forward in 1923) by the humanist and Albert Schweitzer, the 1952 Nobel Peace Prize Laureate, who devoted his life after theology to the service of the poor in the African jungle of Gabon, providing them with free medical care. In his book *Kultur und Ethik*, Schweitzer defines ethics as "infinitely extended responsibility for everything that lives."[18] It is compassion and the experientially grounded but absolute obligation to respect life in all its forms. Schweitzer starts from the premise that human beings are "life that wants to live in the midst of other life that wants to live."[19] Rejecting the common ethics as based on compromise and accepting a trade-off between diverging interests, he argues for an absolute and uncompromising ethics of respect for life. As life as such is sacred, "good" means to preserve, promote, and protect life, "evil" to destroy life and stifle its development. Accordingly, the moral person equally refrains from breaking a branch off a tree, from picking flowers or carelessly destroying insects. Wherever "I sacrifice or hurt life, I am guilty."[20] As it is virtually impossible to observe the strict rules of such ethics, Schweitzer accepts its verdict that he like anybody else who is unable to escape the necessity ("the mysterious and cruel law") to live at the expense of other lives is guilty of murder. Even suicide is not an option, as one's own life is to be respected as much as that of others that may be saved by the self-sacrifice. Nevertheless, all human beings are morally obligated to try everything in their power, to minimize their guilt in the daily practice and reverence of life in all its forms. Schweitzer is well aware of the tragic, even pessimistic undertone of such ethics and its danger of leading to despair. He accepts this as the inevitable consequence of the mystery of life and creation that he is content to venerate in a religious attitude that transcends all institutionalized forms of religion. Schweitzer's ethics is therefore deeply humanistic and the foremost means of expanding our sense of humanity.

A further approach deserves mentioning. It develops in analogy to humanist versions of an ethics of respect, and accepts their account of intrinsic moral worth, as based upon teleological conceptions of moral subjects as ends in themselves. Accordingly, respect for nature would entail the extension of the concept of self to nature, and that is what Freya Mathews has made the center piece of her ethics of the ecological self. On her account, "the universe is itself a self, possessed of its own grand telos and immeasurable intrinsic value."[21] A self, however, has not to be an autonomous and self-conscious person in the traditional sense of the term, and the ethics of the ecological self is non-anthropocentric. What distinguishes a self is its "conatus, its power of self-realization, the will to exist." In its will to exist the self not only realizes itself but reinforces "the cosmic conatus," and helps maintain the ecosystem as a whole. While "animals can contribute to the self-maintenance of the ecosystem without being aware of their unity with it and acting only out of their own narrow self-interest," human beings require "an eco-sensitive culture"[22] that constrains their egoism so as to enable the flourishing and sustaining of all of nature. As Val Plumwood has noted, as long as theories of the expanded self are developed along Arne Naess' deep-ecological conception of identification with nature, they appear to effectively extend human egoism into nature.[23]

## 2. The Boundaries of Nature

One of the problems the ethics of respect for nature has to address, is the apparent difficulty to define and justify the value boundaries of nature, once the morally relevant distinction between humanity and nature has been removed. This had already become an issue in eighteenth-century philosophy, which sought to integrate the insights into nature that the modern science made available. In François Marie Arouet de Voltaire's dialogue between a philosopher and nature,[24] Nature reveals itself to the curious philosopher as "*le grand tout*," the great all and everything outside of which there is nothing, but an elusive God whose faint traces in the fabric of nature are slowly disappearing under the scientific gaze. It seems a consequence of such comprehensive definition of nature when in contemporary environmental ethics, the biocentric restrictions on intrinsic value are increasingly called into question. Keekok Lee has argued for their removal and for the inclusion of "Abiotic Nature," and the extension of moral concern beyond the planet Earth toward "an astronomically bounded ethics."[25] It is questionable whether this approach can accommodate the intuitions that are central to standard conceptions of ethics and morality. Instead it may not only render ethics altogether meaningless, but also distract from and undermine much-needed efforts to save the ravaged planet Earth, and to sustain the common habitat for humans and animals alike.

The ethics of respect for nature is faced with various conceptual difficulties, as the anthropocentric focus of ethics in the Western tradition seems ill suited for the expansion of the locus of moral worth beyond humanity. John Passmore has therefore called for a "new metaphysics"[26] and Val Plumwood has argued that nothing less than a new theory of rationality could provide an adequate basis for a genuine non-anthropocentric theory of intrinsic worth of nature.[27] Others have sought remedies in non-Western traditions, most notably those of indigenous people and holistic philosophies of the East. Especially, Chinese and Taoist thought associated with polar complementarity and pure immanence, instead of a dichotomy between the individual and culture or society, or between the part and the whole have been put forward as promising candidates, for the construal of alternative models of rationality and ethics. It is the longing for a "metaphysics of presence, wholeness, and totality," which seems unavailable within Western traditions that stimulates this search for holistic conceptions of reality, "in which everything exists as part of Ultimate Reality, equal in degrees of being and reality, and fundamentally interrelated as in the luminous image of Indra's net."[28] The favorite Buddhist metaphor of Indra's net has been given prominence by the Hua-yen school and its founder Fa-tsang (643-712 AD). Providing an illustration of the school's doctrine of totality, Fa-tsang covered the walls, floor, and ceiling of a room with mirrors and placed a statue of the Buddha in its center. Holding a reflecting crystal in his hand, he demonstrated how each image of the Buddha reflected from the mirrors was contained in the crystal, and was in turn infinitely reflected back into each mirror.[29] The metaphor stands also in the background of Chu-hsi's (1130-1200 AD) metaphor of the moon, as the simile of the Great Ultimate whose light is scattered upon rivers and lakes but can be seen everywhere as one moon.[30]

Before we embark on what may turn out to be an over-ambitious and doomed enterprise, we should re-assess our own tradition so as to gain better understanding of the magnitude of the task involved. After all, our understanding is largely defined by the "principle of effective history" (*Wirkungsgeschichte*)[31] that affects not only our responses but also the questions we can possibly ask. This is crucial as the concept at the core of the ethics of respect for nature is highly complex and even ambiguous, and has a history that reaches back to the very beginning of Western philosophy and ethics. Nature still looms large in contemporary environmental ethics and is arguably its most contentious conceptual constituent. Nevertheless, Steven Vogel has recently called for discarding it altogether, and for developing an environmental philosophy in its place that "would find its normative foundation in an appeal to self-knowledge and self-recognition" and "not [in] an impossible return to a nature beyond or before human practice." His main point is the claim that nature no longer is – and in fact, it never was – the "wholly other" of humanity. Instead, it is only present as "mediated through language, concepts, world views, and personal and social histories that are particular and contingent; it never appears nor could it appear as it is 'in itself'."[32] I will return to this thesis at the end of this chapter.

Similarly and in a more realistic sense, it has been pointed out that the naturalness of nature has disappeared altogether, and that the ethics of respect for nature has lost its object. "By changing the weather, we make every spot on earth man-made and artificial. We have deprived nature of its independence, and that is fatal to its meaning. Nature's independence *is* its meaning; without it there is nothing but us."[33] Recently, Edward O. Wilson has noted that humans are "the first species to become a geophysical force, altering the Earth's climate, a role previously reserved for tectonics, sun flares, and glacial cycles." Humans are "also the greatest destroyer of life since the ten-kilometer-wide meteorite that landed near Yucatan and ended the Age of

Reptiles sixty-five million years ago."[34] If it is indeed true, as all indicators suggest, that climate change is due to human interference, then original nature in the strict sense with its flora and fauna that exists independently, and prior to human interaction has already been lost and cannot be recovered. While this would affect considerations about nature's inherent worth and the case for preservation (as the maintenance of wilderness, wildlife, and an unspoilt countryside), it would certainly increase the sense of human responsibility for conservation (as the maintenance of resources) and sustainable development.[35]

As we refer to nature in a variety of contexts and in different perspectives, Kate Soper[36] offers the useful distinction between at least three broad categories: the metaphysical, the realist and the "surface" concepts of nature. As it is mainly the metaphysical conception that is the prime focus of the ethics of respect for nature, at least some of its implications need to be considered. We should note from the outset that even this concept cannot be completely separated from the realist, and "surface" notions of nature that stand for the structures and processes of the physical world, and the nature of immediate experience and aesthetic appreciation. I will argue that the ethics of respect for nature is based on a conception of nature that remains elusive, ambiguous, and even muddled as it seeks to integrate conflicting components. Their analysis within the historical and ideological contexts will reveal a structural weakness that undermines the ethics of respect for nature Taylor and others seek to construct upon its foundations.

## 3. Bases of Intrinsic Value

Traditionally, theories of intrinsic value of or in nature, in opposition to humanist or anthropocentric ethics, developed within a religious interpretation of the world, and most notably on the basis of a Christian theology of creation. While any value of creation and of its creatures had its origin in God, and therefore was ontologically prior to human valuation, its inherent worth was relative to God's own nature (being) and his will. Thus different types of theological ethics became available with the stress on rationalistic (Thomas Aquinas; Kant), or voluntaristic (Duns Scotus, William Ockham; Jakob Böhme) interpretations of the divine origin of value. Apparently, the objectivity and universality of value claims derived exclusively from the divine act of valuation implicit in creation as God himself was the supreme value, its source, and measure.

As this solution is no longer available for secular ethics, the search for the inherent worth of nature has shifted in the direction of nature's metaphysical elevation, by means of acquiring attributes that formerly belonged to the Christian God (as it has been famously expressed in Baruch Spinoza's formula *Deus sive natura*), or toward a re-interpretation of values as ontological properties that a specific (affective or intuitive) sense could discover out there "as part of the fabric of the world"[37] that is logically independent of human valuation. Since there could be values without appraisers is counter-intuitive, and requires an explanation of how values are generated, and how humans can know about them. Prominent theories of objective value are associated with the names of Franz Brentano, George E. Moore, Max Scheler and Eduard von Hartmann. The extent to which they are still anthropocentric remains controversial. Their common assumption that it is some kind of natural sense or feeling that detects objective value, and gives rise to moral obligation has been utilized for biocentric versions of environmental ethics. Yet strong claims about nature's values being genuinely independent of human estimation seem to imply that it is neither possible to know about them nor to appreciate (applaud or condemn) them, unless the moral sense is naturalized and ethics turned into a system of involuntary emotional reflexes. Therefore "the argument to value nature 'for its own sake' is self-defeating . . . or it is intended to register the fact that nature is (or should be) valued for other than instrumental purposes."[38]

Robert Elliot has proposed some alternative on the basis of a subjectivist meta-ethics. He seeks to ground intrinsic value in nature without being committed to endorse a narrowly anthropocentric, that is "chauvinist" position, or to reject what he sees as an "unavoidably human perspective."[39] Instead, he derives intrinsic value in nature from a strong conception of "naturalness" that grounds value in the ontological property of "being naturally evolved."[40] While he rejects as "grotesque" the view of "ourselves as simply another species embedded . . . in nature,"[41] he avoids both narrow anthropocentrism and valuation without appraiser.

Provided that moral obligation entails a claim for universal validity, it is not entirely clear how a subjectivist theory of intrinsic value could retain the sense of objectively valid obligation that is associated with moral norms. If the intrinsic value of nature would be plausible only for and within a specific moral community, its significance for global environmental practice and policy would also be seriously diminished.

## 4. Anthropocentrism and Teleology

There is a nagging suspicion that an ethics of respect deriving moral status from factual claims about living organisms as ends in themselves takes too lightly the challenge associated with the naturalistic fallacy. After all it is not only the question how the factual claims could be established and to what extent, but why this fact should confer moral status. Even in Tom Regan's restricted account, inherent value (which in his terminology corresponds to Taylor's inherent worth) is attributed on the grounds that the animals in question are "experiencing subjects of a life," that is "conscious creature(s) having an individual welfare that has importance" to them whatever their "usefulness to others."[42]

As this factual claim has originally been derived from a specifically human experience from where it has been applied to animal behavior, comparability and commensurability of such forms of experience need to be clearly substantiated. It must also be clarified without undue recourse to human experience what it can mean for animals to be experiencing subjects "of a life." At least in humans such experience extends as much into the future as into the past and is, after all, instrumental in the constitution of self-consciousness and a sense of identity over time. Even if one does not share its existential-phenomenological analysis along the lines of Heideggerian ontology, one may wonder whether being a subject "of a life" would have to include some sort of understanding of life's finitude in its natural space between birth and death. It is not easy to see how such a factual claim could be substantiated in animals. In spite of much scientific progress in the study of animals, it is not even clear whether this would require to gain access to their inner experience. What this could mean without sharing a common language or communicative system remains equally dubious. Instead of bringing us closer to the animal world, the claim to know "what it would be for other creatures to consider themselves subjectively as ends" could simply be another case of human arrogance. As we lack "the words to say what it is to be without them,"[43] Peter Strawson may be right that "we must in this matter be content with knowing ourselves."[44] In any case, it is interpretation of animal behavior, interests, and needs from and in a specifically human perspective, and we must be prepared to recognize "limits of our understanding" of other lives.[45]

The issue of teleology, too, requires deeper examination as it has been appropriated as the basis for the extension of respect beyond humanity, and forms the core of a specifically non-anthropocentric ethics. I submit that attempts to rehabilitate teleological thinking and to perceive nature from the perspective of teleology derive their explanatory power from a theory of human action that is constitutive for a sympathetic interpretation of nature. This recalls an Aristotelian philosophy of nature and "natural tendency" that governs organic and un-organic entities alike. Teleology comes into view only through reflection on human intentionality, which is most obvious in self-conscious action. Aristotle rejects earlier attempts by the natural philosophers of ancient Greece since outside human practice the language of intentionality is unintelligible. Goal-oriented processes such as the growth of plants, the behavior of animals presuppose the human perspective, and inner experience of acting. Aristotle reinterprets the sophistic claim of the incompatibility of *physis* and *nomos* as one of complementarity. As the human being is by nature a political being, the dichotomy between nature and humanity is not absolute. If human action can be adequately understood only teleologically and if humanity is part of nature, then all of nature must be understood teleologically. Neither is the distinction between art and nature absolute. On the contrary, nature is not only understood through art, but is itself art. It is this view, which is still remembered even in the midst of the Enlightenment and at the beginning of modern science. In Voltaire's dialogue the philosopher raised the question how it comes that nature looks so raw and chaotic in its mountains, deserts, and oceans, but so purposeful and artful in the realm of plants and animals. Nature answers the astonished philosophers: "My poor child, shall I tell you the truth? People gave me a name, which is inappropriate; they call me nature, and I am all art (*c'est qu'on m'a donné un nom qui ne me convient pas; on m'appelle nature, et je suis tout art*)."[46] This recalls a long tradition, which the founder of the Academy of Florence, Marsilio Ficino (blending Aristotelian and Platonic ideas) summarized as follows: "What is human art? A certain nature, which treats matter from outside. What is nature? A certain art, which forms matter from inside."[47]

"Art," it must be recalled, is in Aristotelian terminology poiesis (*poetike techne*) and represents the second realm of human activity, the skillful production of things according to a plan, and an intention for which things are produced. In praxis (*praktike techne*) the goal is internal, in poiesis it is external. It implies that if nature is art, all natural things are teleologically structured, and therefore the fire strives upward and the stone falls downwards until they reach their appropriate "natural" and that means teleological place. The same applies to plants and animals that grow, develop, and die according to their inherent goals, which express themselves in the various appetites, drives, natural desires, and wants. Nature then, and that includes the whole "universe" whose name, incidentally, still reflects this internal tendency toward its one and single goal (*in unum vertere*), nature is a beautifully ordered, self-contained system (*kosmos*) that has its telos in itself and directs everything toward it. Contemporary Gaia-ecology implicitly depends on a basically Aristotelian world-view and philosophy of nature in spite of its attempt to integrate the results of modern science.

One of the major philosophical problems this world-view raises is clearly noted in Voltaire's dialogue as well. The excited philosopher's question about the ultimate reason for nature's existence, and the final purpose of all its beautiful arrangements and orders remains unanswerable. With remarkable clarity, Nature responds that it does not know its own purpose and what "it all means" (Thomas Nagel), since its goal is in itself. Nature's goal is only its own existence, is in fact self-preservation and nothing else. The German Romantic philosopher-poet Novalis will offer a beautiful metaphor: nature resembles a "mill that grinds itself," without beginning and without end, without purpose and without meaning. This marks the "inversion of teleology" (Robert Spaemann) as the telos of development and growth no longer lies ahead but reverts to its origin. The question: self-preservation for what? It has become meaningless for a specifically non-anthropocentric ethics of respect for nature where the flourishing of one species is as natural as the disappearance of another.

From such a perspective only two options are available, and both have been spelled out in all their consequences in the history of Western philosophy and theology. The first one again is noted by Voltaire's philosopher. With an obvious air of despair he responds to Nature's *Je n'en sais rien*, I have no idea why I am: "Would not non-existence be better than this multitude of existences made in order to be continually dissolved, this crowd of animals born and reproduced in order to devour others and to be devoured, this crowd of sentient beings formed for so many painful sensations, that other crowd of intelligences which so rarely hear reason? What is the good of all that, Nature?"[48] This reaction still resonates with the anxiety of Blaise Pascal who, a century earlier, expressed shock and despair not only about the deafening "silence of the infinite spaces" of the universe, but also about the meaninglessness of a rational being presumably standing at the apex of creation, but facing destruction through the minutest things or simply by a drop of water.[49] This "crown of creation," this receptacle of ultimate truth is nothing but a sewer of filth, an imbecile earthworm and a shifting reef helplessly exposed to hostile elements.

In other words, the first option is nihilism and despair. It may not always express itself in existential anxiety or the courageous anticipation of death in life through which *Dasein* is brought before itself, and called to gain its heroic strength for mastering an otherwise meaningless existence. It can – again in the terminology of Martin Heidegger – simply look the other way and suppress any longing for answers in the average existence of the "man," the anonymous mass of people without face and destiny.

Modern science methodically "brackets" (Edmund Husserl) such questions about the meaning of nature or human existence and instead explores exclusively the realm of causality in the interest of human domination over nature. At the beginning of modernity, when the other, theologically-inspired option was still the dominant background assumption, this methodological self-restriction is expressed in Hugo Grotius's advice for modern science to proceed *etsi deus non daretur*, as if God did not exist.[50]

The second option, namely, is theological. It draws the consequences of an Aristotelian philosophy of nature within a Christian theology of creation and provides nature its ultimate telos, which can only be found outside itself. This revised Aristotelian perspective has been nicely expressed in Thomas Aquinas' consideration that in the case of non-conscious "natural" things the intention is not in them, but in the one who directs them toward their specific ends, like the arrow, which aims at its target not by itself but is directed by the archer.[51] In other words, the reason why nature exists cannot be expected from nature itself, and in that Voltaire's Nature gives an honest and truthful answer when she utters her *Je n'en sais rien.* At the same time, she has still preserved the old Christian vision of a purpose and a goal beyond itself and directs the anxious philosopher toward its own maker: "Go and ask him who made me!" Voltaire's nature then stands at the threshold that separates the Christian world-view from the modern, scientific one and a Christian-based ethics of ontological value from a secular ethics of respect for nature.

## 5. Nature Between Deification and Desacralization

The transformation of Aristotelian teleology in Christian medieval thought, encapsulated ideological possibilities whose explosive force would be released once the personal God became elusive, and a mere philosophical concept whose main significance derived from its function in the rational interpretation of the universe. On the assumption that the finality of nature requires consciousness that cannot be found in nature itself, but refers to divine intention and consciousness external to nature, nature is effectively seen as the product of divine art designed to follow its own causal laws and only indirectly revealing its maker. Nature had become that vast machine whose blueprint can be read like a book which written in numbers and geometrical figures. As a contemporary dictionary states: "this vast Machine of the Universe, the wise Production of Almighty God, consisting of a great number of lesser Machines, every one of which is adjusted by the same Wisdom in Number, Weight, and Measure."[52] For an age that admired the construction of mechanical clocks and for want of anything better nature is itself thought to resemble a divinely constructed clock whose precision of movement the human analogue seeks to imitate, and whose study would direct the diligent researcher to its true designer.

Or so it seems. The metaphor of the book of nature is not only found in Galileo Galilei's famous remarks, but was a widely accepted notion whose significance derived from its implicit claim for equal dignity with the book of revelation. The tradition dates back to Clement of Alexandria, Origen, Augustine and other Church Fathers and has been revived in the fifteenth century by Raymond of Sabonde's *Theologia Naturalis* (1436 AD): *Libri duo dati sunt a deo scilicet liber naturae et liber sacrae scripturae*. Although both books have God as their author, the book of nature can be understood through reason alone and independently of the insights gained through revelation and its authoritative interpretation by the Church. As soon as the original complementarity is replaced by an indifferent dualism, nature alone provides the means for its adequate understanding and the "hypothesis of God" has become obsolete. The early development of science illustrates the process of its gradual alienation from the religious world-view toward the latter's outright rejection.

Apparently, the opposing views on nature implied in contemporary environmental philosophy, and in the natural sciences have been anticipated in early modern development when the apparently barren Cartesian dualism engendered nature's divinization and its utter desacralization. Nature could be equally metaphysically devalued and turned into a blind automaton whose structures would become the object of scientific curiosity, as it could be metaphysically elevated and transformed into a goddess endowed with the attributes that originally belonged to the Christian God. As Carl L. Becker observes: "The eighteenth century . . . did not cease to bow down and worship; it only gave another form and a new name to the object of worship: it deified Nature and denatured God."[53] It was this goddess Nature whom the baron Paul Henri Thiry d'Holbach addressed at the end of his *Système de la Nature* (1770 AD) as "sovereign of all beings" to whom "belong the praises and the homage of the earth" and who alone could bestow true happiness on those obeying her commands. The seventeenth-century movement of physico-theology tried to steer a middle course and to utilize science for the renewed search for the traces of the Christian creator God in his creation. Unfortunately, the great scientific erudition and scholarship it invested for providing proof of God's existence from design concerned a lost cause as it was "proof" only for those who believed anyway.[54]

Both metaphors, that of nature as a vast machine (clock) and of a book, clearly indicate that it is not nature as the absolute other of humanity, pure nature, they are concerned with but nature as a cultural product and to a certain extent a human construct. Nature as nothing-but-nature, as "mere" nature, is only available through a complex and deliberate process of reduction and abstraction, as is indeed the case in the methodological reductionism of modern science. This realist nature is generated through a process of abstracting from what is otherwise the most "natural" human perspective through which nature reveals itself in its meaning and significance for human life and self-understanding.

In this regard, the sciences of nature are themselves a product of culture and human history. The personalistic disposition is ontologically prior to the naturalistic and the understanding of nature through the so-called "natural sciences" entails a reductionistic perspective. Vogel's "societal constructivism" is at least partially correct as nature is indeed "always already interpreted" and "we have no access to nature in itself."[55] I also concur with him that "we have to resist the temptation to hypostatize, glorify, reify, or deify" nature once again.[56] Yet this does not make "pointless and indeed conceptually incoherent" to distinguish between the human and the natural world,[57] especially as the human and the natural world stand in dialectic relationship to each other. The "I" at the center of my world can only be itself over against the "other" that I am not. This dialectic can be illustrated with reference to the human body as the locus of our "being-in-the-world." The body (*Körper*) as natural organism is part and parcel of nature and the natural world around us. In this regard, it is in principle not different from any other organism, and this is the main and sometimes the sole perspective of the biocentric world-view. It is also the object of the scientific perspective I assume whenever I examine my body for defects and subject it some sort of maintenance or repair; it is this body that takes center stage within the Western paradigm of medicine.

As the body is also and above all the locus of *my* appearance in the world, it is always more than the object of scientific exploration. It is *Leib* (cognate of life) or body in the emphatic and specifically human sense, and defines what Helmuth Plessner called the "eccentric position" of human beings, which includes the having-of-a-body and the being-a-body. As humans, we not only "have" a body (*Körper*), but also *we* "are" body (*Leib*); *we* exist corporally in an emphatic sense. The body stands for the presence of nature in ourselves as natural beings and for the distance that separates *us* from nature.[58]

It follows that interpretations of nature are not only human self-interpretations, but truly interpretations of nature. Nature proves its distinctness, above all, in our failing interpretations. As Kate Soper remarks succinctly: "it is not language that has a hole in its ozone layer; and the 'real' thing continues to be polluted and degraded even as we refine our deconstructive insights at the level of the signifier."[59] In addition, it is also we who define the goals of science, and no one else but we are responsible for the direction of research and the use we make of its results. Insofar as the inevitability of the scientific development (*Sachzwänge*) that gradually escapes human controls becomes reality, a reminder of the cultural origin of science is the call for assuming responsibility for our products before they enslave their creator and spoil nature beyond repair.

## 6. The Ineliminable Human Perspective

To conclude, in view of the dialectic of nature and humanity, the issue of anthropocentrism becomes ambiguous and highly complex. On the one hand, the charge of anthropocentrism has been taken to imply an undue focus on human beings and their unique role in nature, at the expense of other beings or even natural things. It also implies that the attempt to understand and to value nature from a human perspective is neither reasonable nor justifiable. In this regard, the Aristotelian perspective (which understood nature in analogy to human purposeful action) is effectively rejected.

Yet when, in the attempt to overcome such narrow anthropocentrism, nature itself is understood as a purposeful system that provides the basis for teleology in all living organisms and even beyond, such an interpretation presupposes the human perspective instead of abolishing it. I take it as a true Aristotelian insight that we simply cannot know anything about purpose and goals before we have encountered and experienced them in our own actions. And from here we extrapolate ends into the world around us. It is in self-conscious beings like ourselves where the language of purpose and intention has its origin and from where it can meaningfully and analogically be applied to non-human nature. This is what Bernard Williams has called "the ineliminable human perspective." If this should still be regarded as anthropocentrism, it is certainly not of the narrow variety that implies that all our value concerns and attitudes are simply and necessarily directed to human interests.[60] While human beings are beings of and within nature, they also signify the "point" where nature opens her eyes and gains consciousness of itself. This idea has been given poetic expression in Johann Wolfgang von Goethe's statement (1782 AD): "Nature is without language and speech but creates tongues and hearts by which she feels and talks."[61] Friedrich Nietzsche stretches this idea even further when he sees in humanity the mirror nature holds up so as to find its own meaningfulness that is no longer external as the for ever elusive goal of its evolutionary urge but eternally present within.[62]

Anthropocentrism plays a problematic role in all attempts to understand nature as the "extended self" and to interpret its operations in terms of teleological self-realization. In as far as this requires "a deep identification of individuals with all life forms,"[63] such individualization of nature depends on a philosophy of subjectivity that cannot deny its origin in human self-consciousness and reflexivity. To the extent that this philosophy provides the vocabulary and the attributes for nature, it anthropomorphizes nature and ultimately aims at creating nature truly in the image of human subjectivity.

At the same time, it unintentionally but factually results in the re-divinization and re-deification of nature, since this approach is no longer content with the methodical atheism of modern science, but interprets nature as the locus (repository) of purpose, value, and worth from where moral obligations can be derived. Nature indeed is turned into a lawgiver who has prescribed the moral norms in the eternal tablets of rocks and stones and in the natural order that defines all biotic systems. *Deus sive Natura* once again have become interchangeable perspectives.

Bernard Williams noted the claim "that human beings should not see the world as simply theirs to control, is often thought to have a religious origin."[64] Traditionally, only religions draw absolute lines that separate the permissible from the impermissible, the profane from the holy or sacred. Etymologically, it will be recalled, the profane is all what is outside or in front of the temple precinct (*fanum*) and what has been "set off, restricted" (*sacer*) from the sacred. Apparently, for a biocentric ethics of respect, human encroachment on nature and even more so its exploitation and destruction breaks down this final taboo and destroys the last remnants of the holy in a secular world.

With the removal of the last barriers nothing remains that would be off-limits and, in principle, everything is at our disposal, including our genome and our own species nature. In the romantic language of Nietzsche, this means with the death of God everything is allowed; and "God" stands for both the Christian creator God and deified Nature.[65] In Wilson's language of naturalist reductionism it means that we have entered the phase of "volitional" evolution: "Homo sapiens is the first truly free species" and "about to decommission natural selection." As there is "no genetic destiny outside our free will" all that is left for us is to "look deep within ourselves and decide what we wish to become." And indeed we have already begun to mold nature in our own image. The prospect of post- or trans-humanist evolution is no longer confined to internet chat rooms but has become a serious issue in biogenetics.[66]

If this is anthropocentrism and wrong, the charge may be human hubris and a complete misjudgment of human knowledge and foresight. But it may not be a call for a return to a religious ethics of respect for nature that sets ultimate limits to human action from outside.

Against this backdrop, the ethics of respect for nature may indeed have to be understood as a return to a non-secular ethics of ultimate concern that wishes to erect new barriers to human hubris by identifying absolute values beyond the disposition (*Verfügbarkeit*) of humanity. If this is thought possible through nature, nature in fact takes over the place traditionally reserved for God.

In environmental debate then the charge of anthropocentrism takes on an air of ambiguity. It may simply signal a sense of loss of the sacred in a thoroughly secular world and in humanist ethics but otherwise unite in the construction of an environmental ethics that extends the scope of moral concern beyond humans and seeks to integrate an aesthetic conception of nature that gives due recognition to its beauty and non-instrumental worth. In its strong version, however, the charge of anthropocentrism would be the call for the re-enchantment of nature and its divinization without transcendence. Even in the case that nature were to be installed in God's place it would remain a "mortal god" (Thomas Hobbes) whose fate is entirely in the hands of humanity.

## NOTES

1. Paul Taylor, *Respect for Nature: A Theory of Environmental Ethics* (Princeton, N.J.: Princeton University Press, 1986).

2. Immanuel Kant, "*Grundlegung zur Metaphysik der Sitten*," *Immanuel Kants Schriften* (Akademie-Ausgabe) (Berlin: de Gruyter, 1903), Vol. 4, p. 401.

3. Allen Wood, *Kant's Ethical Thought* (Cambridge, UK: Cambridge University Press, 1999), p. 46.

4. Stephen L. Darwall, "Two Kinds of Respect," *Ethics*, 88:1 (1977), pp. 36□49, p. 44, p. 45. See also, Carl Cranor, "Toward a Theory of Respect for Persons," *American Philosophical Quarterly*, 12 (1975), pp. 303□319.

5. Martha C. Nussbaum, *The Therapy of Desire: Theory and Practice in Hellenistic Ethics* (Princeton, N.J.: Princeton University Press, 1994), p. 12.

6. *Ibid.*, p. 325.

7. *Ibid.*, p. 326.

8. Theo Kobusch, *Die Entdeckung der Person: Metaphysik der Freiheit und modernes Menschenbild* (Darmstadt: Wissenschaftliche Buchgesellschaft, 1997).

9. Immanuel Kant, "*Metaphysik der Sitten*," *Immanuel Kants Schriften* (Akademie-Ausgabe) (Berlin: de Gruyter, 1907), Vol. 6, p. 392.

10. Kant, "*Grundlegung zur Metaphysik der Sitten*," Vol. 4, p. 437.

11. Tom Regan, "The Case for Animal Rights," *Ethics: History, Theory, and Contemporary Issues*, eds. Steven M. Cahn and Peter Markie (New York: Oxford University Press, 1998), pp. 821□829, p. 827.

12. Paul Taylor, *Respect for Nature*, p. 122.

13. *Ibid.*, p. 15.

14. *Ibid.*, p. 178.

15. *Ibid.*, p. 84.

16. *Ibid.*, p. 99.

17. *Ibid.*, p. 100.

18. Albert Schweitzer, *Kultur und Ethik* (München: C. H. Beck, 1923), esp. pp. 237□261 (Die Ethik der Ehrfurcht vor dem Leben), p. 241.

19. *Ibid.*, p. 239.

20. *Ibid.*, p. 257.

21. Freya Mathews, "Value in Nature and Meaning in Life," *Environmental Ethics*, ed. Robert Elliot (Oxford: Oxford University Press, 1995), pp. 142□154, p. 147. See also, *The Ecological Self* (London: Routledge, 1991).

22. *Ibid.*, p. 148, p. 149.

23. Val Plumwood, "Nature, Self, and Gender: Feminism, Environmental Philosophy, and the Critique of Rationalism," *Environmental Ethics*, ed. Robert Elliot (Oxford: Oxford University Press, 1995), pp. 155□164, p. 161.

24. François Marie Arouet de Voltaire, *Dictionnaire Philosophique*, *Oeuvres Complètes de Voltaire* (Kehl: l'Imprimerie de la Société littéraire-typographique, 1784□1789), Vol. 61; see also, *Dictionnaire de la pensée de Voltaire par lui-même*, ed. André Versaille (Bruxelles: Edtitions Complexe, 1994), pp. 917□918, p. 917.

25. Keekok Lee, "Awe and Humility: Intrinsic Value in Nature. Beyond an Earthbound Environmental Ethics," *Philosophy and the Natural Environment*, eds. Robin Attfield and Andrew Belsey (Cambridge: Cambridge University Press, 1994), Royal Institute of Philosophy Supplement Vol. 36.

26. John Passmore, "Attitudes to Nature," *Environmental Ethics*, ed. Robert Elliot (Oxford: Oxford University Press, 1995), pp. 129–141, p. 138.

27. Val Plumwood, *Environmental Culture: The Ecological Crisis of Reason* (London: Routledge, 2002).

28. Linda E. Olds, "Chinese Metaphors of Interrelatedness: Re-Imaging Body, Nature, and the Feminine," *Contemporary Philosophy*, 13:8 (1991) pp. 16–22, p. 21.

29. David L. Hall and Roger T. Ames, *Thinking Through Confucius* (Albany: State University of New York Press, 1987), p. 239.

30. Cf. Wing-tsit Chan, ed. and trans., *A Source Book in Chinese Philosophy* (Princeton, N. J.: Princeton University Press, 1969), pp. 412, 638–639.

31. Hans-Georg Gadamer, *Wahrheit und Methode* (Tübingen: Mohr, 1960).

32. Steven Vogel, "Environmental Philosophy after the End of Nature," *Environmental Ethics*, 24 (2002), pp. 23–39, p. 32, p. 24.

33. Bill McKibben, *The End of Nature* (New York: Anchor, 1989), p. 58.

34. Edward O. Willson, *Consilience. The Unity of Knowledge* (New York: A. Knopf, 1998), pp. 275–277.

35. John Passmore, *Man's Responsibility for Nature* (London: Duckworth, second ed. 1980), p. 73.

36. Kate Soper, *What is Nature? Culture, Politics and the non-Human* (Oxford: Basil Blackwell, 1995), pp. 155–156.

37. John L. Mackie, *Ethics. Inventing Right and Wrong* (Harmondsworth: Penguin Books, 1997).

38. Kate Soper, *What is Nature?* p. 255.

39. Robert Elliot, "Ecology and the Ethics of Environmental Restoration," *Philosophy and the Natural Environment*, eds. Robin Attfield and Andrew Belsey (Cambridge: Cambridge University Press, 1994), pp. 31–43, p. 33; see also, Robert Elliot, *Faking Nature: The Ethics of Environmental Restoration* (London: Routledge, 1997), further, "Instrumental Value in Nature as a Basis for the Intrinsic Value of Nature as a Whole," *Environmental Ethics*, 27 (2005), pp. 43–56.

40. *Ibid.*, p. 36, p. 37.

41. *Ibid.*, p. 42.

42. Tom Regan, "The Case for Animal Rights," *Ethics: History, Theory, and Contemporary Issues*, eds. Steven M. Cahn and Peter Markie (New York: Oxford University Press, 1998), pp. 821–829, p. 827.

43. Kate Soper, *What is Nature?* p. 173 and p. 171.

44. Peter Strawson, *Bounds of Sense* (London: Methuen, 1996), p. 273.

45. Kate Soper, *What is Nature?* p. 173.

46. Dictionnaire de la pensée de Voltaire, p. 917.

47. Marsilio Ficino, *Theologia Platonica de immortalitate animorum: XVIII libris comprehensa* (Hildesheim: Olms, 1975, reprint of 1559 edition), Part IV, Sect.1.

48. Voltaire, *The Philosophical Dictionary*, trans. H.I. Woolf (New York: Knopf, 1924). http://history.hanover.edu/texts/voltaire/volnatur.html

49. Blaise Pascal, *Pensèes* (Paris: Le Livre de Poche, 1972), No. 206, 194, 347.

50. Hugo Grotius, *De Iure Belli ac Pacis* (Paris, 1625).

51. Sancti Thomae Aquinatis Summa theologiae: cura fratrum eiusdem ordinis (Matriti : La editorial catolica, 1961), I, q. 2, a. 3.

52. John Harris, *Lexicon Technicum or an Universal English Dictionary of Arts and Sciences* (London: D. Brown, 1702), quoted by Jürgen Mittelstrass, "Leben mit der Natur," *Über Natur*, ed. Oswald Schwemmer (Frankfurt: Klostermann, 1991), pp. 37–62, p. 39.

53. Carl L. Becker, *The Declaration of Independence* (New York: Alfred Knopf, 1966), p. 51.

54. See Gerhold K. Becker, "The Divinization of Nature in Early Modern Thought," *The Invention of Nature*, eds. Thomas Bargatzky and Rolf Kuschel (Frankfurt: Lang, 1994), pp. 47–61.

55. Steven Vogel, "Environmental Philosophy After the End of Nature," *Environmental Ethics,* 24 (2002), pp. 23–39, p. 35.

56. *Ibid.*, p.39.

57. *Ibid.*, p.32.

58. Bernhard Waldenfels, *Das leibliche Selbst* (Frankfurt: Suhrkamp, 2000), p. 252–253.

59. Kate Soper, *What is Nature?* p. 151.

60. Bernard Williams, "Must a Concern for the Environment Be Centered on Human Beings?" *Making Sense of Humanity* (Cambridge: Cambridge University Press, 1995), p. 235.

61. Johann Wolfgang von Goethe, "Die Natur. Fragment" (1782), *Goethes Werke in Zehn Bänden*, ed. Reinhard Buchwald (Hamburg: Standard-Verlag, 1957), Vol. 4, p. 269. (*Sie [die Natur] hat keine Sprache noch Rede, aber sie schafft Zungen und Herzen, durch die sie fühlt und spricht).*

62. Friedrich Nietzsche, *Sämtliche Werke. Studienausgabein 15 Bänden*, eds. Giorgio Colli und Mazzino Montinari (München: Deutsche Taschenbuchverlag, 1980), Vol. 1, p. 378.

63. Arne Naess, *Ecology, Community and Lifestyle* (Cambridge: Cambridge University Press, 1989), p. 38.

64. Bernard Williams, "Must a Concern for the Environment," p. 238.

65. See also, Elizabeth Anscombe, "Modern Moral Philosophy," *Philosophy*, 33 (1958), pp. 1–19.

66. See Allen Buchanan, Dan Brock, Norman Daniels and Daniel Wikler, *From Chance to Choice: Genetics and Justice* (Cambridge: Cambridge University Press, 2000).

# Three

# ENVIRONMENTAL ETHICS: AN AESTHETIC APPROACH

Ingmar Persson

## 1. Introduction

As a branch of ethics, environmental ethics is anomalous. It is not centered on the sort of individuals that other branches of ethics, in particular its closest kin animal ethics, are centered on. Environmental ethics does not assign to these individuals, for instance, animals, a moral status or standing in the sense that our treatment of them can be morally right and wrong, regardless of effects on anything external to them. In other branches of ethics, you can be said to treat a collective morally rightly or wrongly, for instance, a coach can be said to treat his/her team wrongly or badly. But this is reducible to treating individual members of the team wrongly or badly. In contrast, as acting wrongly in environmental ethics is a matter of treating individuals that compose the environment morally wrongly.

For instance, you act nothing wrong in environmental ethics if you painfully kill individuals of some prolific species, like the brown rat, for no good reason. Even though on most moral views animals like rats are among the individuals that have moral status, so that painfully killing them for no good reason would be wrong. And they are doubtless among those individuals that compose the environment with which environmental ethics is concerned. As we will see when an adequate criterion of moral standing has been laid out, you can act environmentally wrongly without harming or doing wrong to any individuals that have moral standing. Consequently, it appears that acting wrongly to beings with moral standing is unnecessary and insufficient for acting wrongly in environmental ethics. In the end, these considerations will lead me to conclude that environmental "ethics" is not, strictly speaking, a branch of ethics. Instead, it is an *aesthetic* concern for the environment which is akin to the concern for the preservation of old buildings.

## 2. What is Moral Status?

Not all theorists assign moral status to non-human animals. There are *anthropocentric* views according to which human beings have a special moral status in virtue of being human beings, so that actions that affect them have a special moral weight – or, more radically, so that actions can be morally right or wrong, only when they affect them. Such an understanding of moral status is sometimes thought to be sufficient for the purposes of environmental ethics.[1]

However, I side with people like Peter Singer who contend that anthropocentrism is "speciesist," that is, unjustifiably discriminatory like racism and sexism.[2] My sympathy lies with the view – it might be called *conativism* – that something has moral status in virtue of being in possession of desires. As I claim, without being able to give sufficient argument here, that for something, S, to have moral status now, there must be things that can be good or bad for S at some time or other. Something is good for S in so far as it satisfies desires that S has, and it is bad for S in so far as it frustrates S's desires. (The relevant desires must meet several conditions, such as not being based upon cognitive mistakes.[3]) Pain is bad for S in so far as S is averse to pain, and pleasure is good for S in so far as S desires to experience it. But a subject can have desires only if it has consciousness, since to have a desire, is to have some thought or sensation which motivates you.[4] It is logically possible that a subject has consciousness, but not desires. Since organisms on earth have consciousness only because of its survival value, this possibility is in all likelihood not yet realized.

The subject matter of morality is, reasonably, the treatment of entities that have moral status in the sense that actions can be assessed, as morally right or wrong, in virtue of their effects on them only. I have proposed that something now has moral status, if and only if, something can be good or bad for it, at some time or other because it is then endowed with desires that can be satisfied and frustrated. (Note that this claim allows for the possibility that something can have moral status now, and without having desires now, say, that a fetus has moral status now because it has the potential to acquire desire in the future.) When I said that environmental ethics is anomalous in not being centered on individuals, I meant that acting wrongly to individuals who have moral status in virtue of having desires at some time, is insufficient and unnecessary for acting wrongly in environmental ethics. This provides a reason for saying that so-called environmental "ethics" is not a branch of ethics, for the subject matter of ethics or morality is, to repeat, the treatment of entities with moral status.

Since conativism is a premise of this argument, let me try to prop up this premise by examining the concept of value a bit closer. It is a formal property of values that they admit of *grades*, that something can be more or less valuable, more or less good or bad. An adequacy condition for an analysis of value is that it can make sense of this formal property. Conativism is well-equipped to do so for desires can also be more or less strong. Typically, our desire to stay alive is stronger than our desires for most other things, say, to experience what it is like to fall freely. This is why we do not jump off skyscrapers to have this experience, as staying alive is of greater value for us than having this experience. Since some animals have stronger and more diversified desires than other animals, it is morally worse to frustrate their desires by killing them. In this way, conativism can explain why it is worse to kill a normal human being than non-human animals, even the more highly developed animals.

But, in contrast to anthropocentrism, conativism does not imply that human beings have a higher moral status by virtue of having the property of being human beings or of belonging to the species *homo sapiens*. As I have suggested, this would be unjustifiable discrimination. Most would agree that it is unjust discrimination to hold that the desires or interests of some human beings have less moral weight because they belong to a particular race. We can make this vivid by imagining that we ourselves were members of that race. We would then find it unacceptable that our interests count for less. But, as the Golden Rule has it, you should do onto others as you want them to do onto you. So, we should not treat people of other races in ways that we could not accept were we members of those races.

Genetically speaking, the difference between belonging to different races and belonging to different species is only a matter of degree. Sometimes it is indeterminate whether the genetic differences are large enough for animals to belong to different species, instead of merely to different races. (If all the intermediate stages between human beings and chimpanzees had survived, we would countenance individuals as regards which it was indeterminate whether they were humans too.) Therefore, we should expect that we could run against speciesism, the same thought-experiment as I just ran against racism.

We can imagine there to be individuals – perhaps on another planet – who are just as intelligent as we are, and developed a civilization just as advanced, scientifically and artistically, as our civilization is, but who belong to another species than *homo sapiens*, because they are genetically so different from us. (Inter-fertility is a common criterion of membership of the same species.) To the naked eye, these individuals might even look indistinguishable from us. If we were to imagine ourselves to be members of that species, we would find it just as unacceptable that our interests count for less – for instance, that we be hunted for food or enslaved – as we found this to be when we imagined ourselves to be members of a different race.[5]

Consequently, I reject the anthropocentric construal of moral status as an instance of unjustifiable discrimination. In the end, I will arrive at an anthropocentric conception of environmental concern, but that is because I do not regard environmental concern as a genuine branch of ethics.

There are writers who take a broader view of the things that have moral standing than conativism does, just as conativism takes a broader view of the things that have moral standing than anthropocentrism. Some of these writers maintain that all biologically living organisms, including plants and microorganisms, have this status in virtue of being biologically alive.[6] We may call this view *vitalism*. Other views go still further and take it that even inanimate things, like rocks and rivers, which help to compose ecological systems, have moral status.[7] Vitalism and these omnibus views, both face difficulties in accounting for the formal feature of value I mentioned above, namely, that value is gradual. I will use vitalism to illustrate this point.

The vitalistic view that things have moral standing in virtue of having biological life can be taken as stating a condition which is necessary and sufficient for moral standing or a condition that is only sufficient. Suppose it is taken in the first way. According to this view, the state of feeling pain is not anything that we have moral reason to abolish or alleviate for. So, it is presumably not anything that can be *intrinsically* bad for the subject feeling it. It is presumably only *instrumentally* or *extrinsically* bad for the subject to the degree that it, or the bodily damage which causes it, is detrimental to the biological life of the subject. This is strongly counter-intuitive: pain is surely bad in itself if it is disliked for its own sake and induces suffering. This fact is enough to provide a moral reason to alleviate the pain, irrespective of the effects the pain may have on the subject's biological life.[8] Imagine, as is surely conceivable, that something which is not a biological organism, for example, an advanced computer, felt a pain sensation that it thoroughly disliked. That would surely be intrinsically bad for it, and therefore something we had a moral reason to abolish, though the computer has no biological life that can be affected by the pain.

Suppose, then, that the idea is only that being biologically alive is a sufficient condition, alongside other sufficient conditions, like the one advocated by conativism, for the possession of moral status. We then face the problem of having to weigh two sources of value that appear to be incommensurable. For instance, what is worse, causing an animal severe pain or destroying the life of a plant? Or how much worse does the damage you do to an animal's life if this is accompanied by an excruciating pain? There are not even roughest answers for these questions.

There are problems even if we confine ourselves exclusively to the value allegedly provided by biological life. Most of us do not want to say that everything that is biologically alive has the same value, for example, that one cell which helps to constitute a human organism has the same value as the human organism constituted. The only reasonable sense in which organisms can be said to be alive to different degrees appears to be that they can be more or less complex. A possible suggestion might then seem to be that the value of a complex organism consists in the sum of the value of all the living cells composing it. This proposal has the merit of making the value of a human organism enormously higher than the value of any of the cells composing it. But it has the counter-intuitive corollary that the value of the life human being decreases if it loses a bit of its living body, say, a finger or the appendix. This is again strongly counter-intuitive. I cannot find any plausible way in which the vitalistic view that biological life is the ground for moral status can account for the gradations of value that we intuitively like to make.

But whichever of these individualistic views of what has moral standing you adopt – antropocentrism, conativism, vitalism, or some omnibus view – you must concede that you can harm or damage some organisms or natural things without harming the environment, without acting wrongly in environmental ethics. This is so, even though such harming or damaging does not produce any beneficial effects for other morally significant constituents of the environment which outweigh the bad ones. The killing of some prolific animals, like the brown rat, for no good reason might be irrelevant from the point of view of environmental ethics, though it is morally wrong.

This is the reason, I think, why *holistic* views in environmental ethics are plausible. The environment might reasonably as a whole which is something over and above the individual parts that compose it. If this whole has intrinsic value and moral standing, it would be intelligible how you could damage or even destroy some parts of it without affecting the value of the whole, for some parts might not be necessary for the intrinsic value of the whole. Holistic views claim precisely this. They deny that acting wrongly in environmental ethics is a matter of acting wrongly to some individual things with moral status that constitute the environment, and this branch of ethics is not centered on the individuals that other branches of ethics are centered on. They claim instead that what has intrinsic value and moral status are whole ecological systems, and that individuals are morally relevant only in proportion to their contribution to these ecosystems.

For instance, a spokesman of this kind of holistic ethics, J. Baird Callicott writes: "the individual drops out of the picture altogether, leaving only the biotic community as the object of respect and moral considerability." He quotes and with the approval of Aldo Leopold's (consequentialist) criterion of moral rightness and wrongness: "A thing is right when it tends to preserve the integrity, stability, and beauty of the biotic community; it is wrong when it tends otherwise."[9] (Callicott insists that the emergence of this holistic morality, "ecocentrism," "does not submerge and override smaller, more primitive human communities and their moral codes."[10] Instead, "our obligations to family and friends – and to human rights and human welfare generally – come first."[11] But Callicott's ecocentrism apparently provides no room for obligations to non-human animals. This leaves Callicott open to the charge of anthropocentric speciesism, I think. Without this exception in favor of human beings, Callicott's ecocentrism would however harbor an implication that few could accept, namely that human beings are morally crucial only to the degree that they make a positive contribution to ecosystems.)

Since I have declared my allegiance to conativism, it follows that I must reject these holistic views, for wholes like ecosystems do not have consciousness and desires (since they consist in living organisms interacting with non-living elements). My view is that the environment can have moral importance only to the extent that it consists of sentient beings which have moral status. Otherwise, we will come up against the problems that were illustrated in connection with vitalism. As will transpire in due course, however, I allow that constituents of the environment that lack moral status can still have intrinsic value, value in themselves. As for reasons that will emerge, I do not think that the possession of intrinsic value is sufficient for possession of moral standing, though it is necessary.

### 3. Environmental Ethics is Essentially Consequentialist

If it is true that you can gratuitously harm individual beings with moral standing that help to constitute the environment without acting environmentally wrongly, it follows that some well-known moral principles do not hold in environmental ethics. These principles include, first, the *Kantian* principle that you must act so that you never treat anybody *merely as a means*, but always simultaneously as an end.[12] (Kant probably restricted his principle to persons or rational beings, but let us bracket this restriction.) They include, secondly, the *Pauline* principle that you must not do *harm or evil* so that good comes out of it.[13] These principles are not equivalent. I may treat sentient beings merely as a means without harming them. For instance, in tailing you, I may hide behind other pedestrians, without interfering with them in any way. Although I am using these pedestrians merely as means, merely as opaque objects behind which I can hide, I do not seem to act wrongly toward them, since I do not harm them in any way. With respect to this example, the Pauline principle fares better than the Kantian principle.

On the other hand, many would hold that it is not wrong to harm someone in the process of doing good, if you do not use the victim as a means. For instance, they would maintain that it is not wrong to deflect a deadly threat, like the well-known runaway trolley, so that it kills one instead of many because the killing of the one is not here a means to the saving of the many. In contrast, it would not be permissible to harm one person – say, to push him or her in front of the trolley – as a means of saving many. On the basis of these considerations, it might be suggested that the most plausible principle is an amalgamation of the Kantian and the Pauline principles. Such a principle could lay down that it is wrong to use sentient beings, in a way that harms them, as a means even to a greater good.

But whatever the merits of such a principle in other departments of ethics, it cannot have any place in environmental ethics. Environmental ethics which is often called *consequentialist.* It allows you to treat individual beings with moral status any way you like, as long as the environmental consequences overall are good enough. As far as environmental ethics is concerned, you have to justify in killing one member of an endangered species, for instance, a snow leopard, as a means of saving several of its conspecifics. Even more obviously, you are allowed to kill off large numbers of rats and feral cats to protect a few members of an endangered species. The end of environmental ethics is thought to justify harming as a means, which puts no constraints on means as long as the end is good enough, it is therefore *consequentialist.*

## 4. The End of Environmental Ethics and Other Consequentialist Ends

What, then, is the end of environmental ethics? I do not think that it can be precisely defined. But, roughly, it is something like *the preservation of a pristine natural environment of global extension.* By calling this environment "natural," I do not mean to imply that it must be untouched by the human hand (there is hardly any such environment left on the surface of the earth). I mean to contrast it with an environment, which is more or less *wholly* constructed by human agents, like a city. To bc pristine, it must contain a rich biological diversity, a manifold of life forms. This end is utterly different from the ends of familiar consequentialist moral theories.

The best known of these theories is probably *utilitarianism.* There are many forms of utilitarianism, but for our purposes, we may take the utilitarian end roughly to be that there be as many as possible beings with lives, that are as good for them as possible. Otherwise, the utilitarian end is to produce the greatest net balance of what is good for beings over what is bad for them. It is fairly obvious that we do not know that preserving the environment will realize this utilitarian end. On the contrary, we know that the natural order involves a lot of suffering. Life involves so much suffering as Buddhists and Arthur Schopenhauer, claim that it is *altogether* suffering.[14]

Many of us believe this to be an exaggeration, but life on earth involves so much suffering that were we accurately to execute the utilitarian calculation – which is beyond our competence – the net-balance would not come out as negative. It is even harder to be sure that we could not boost the utilitarian sum – increase animal welfare – by interfering radically with the environment, say, reduce the number of killings performed by predators. So, we cannot defend the pursuit of the environmental ethical end of preserving the environment, by claiming that it coincides with the utilitarian end of maximizing the net-balance of what is good for beings over what is bad for them. This is quite compatible with the pursuit of the environmental end, being better from a utilitarian perspective than environmental destruction in the service of, especially in short-term, human interests.

The natural order involves an extraordinary amount of *injustice* (a moral aspect which utilitarianism is renowned for ignoring). Some sentient creatures are born better equipped to lead good lives than their fellow creatures. These stronger individuals tend to eliminate the weaker, in the Darwinian struggle for reproductive success. It is unjust that, through no fault or choice of the inferiors, some are born with smaller chances to lead good lives and, as a result, fail to lead such lives. And the natural order is glaringly unequal and unjust. Consequently, from the point of view of utilitarian maximization and justice, the natural order is far from morally ideal, and it may even be morally unjustifiable. It appears that we cannot justify the end of environmental ethics, to preserve a pristine environment, by maintaining that it is conducive to welfare-maximization or to a just distribution of welfare over the beings with moral status that help to constitute the environment. The end of environmental ethics is different from the ends of consequentialist moralities.

## 5. The Environmental End Aesthetic

How could we justify the end of environmental ethics? I believe that this end represents an *aesthetic* ideal instead of a moral one. It is helpful to compare the aim of preserving old buildings. We want to preserve these buildings not because it is good for them, instead they benefit from being preserved. Since buildings do not have consciousness and desires nothing can be good or bad for them, nothing can benefit or harm them. We preserve these buildings because of their beauty. But we also preserve them for what we can learn from them about our past. Analogously, I think it is desirable to preserve the natural environment, first, because it is beautiful, magnificent, and awesome; and, secondly, because it can tell us things about the evolution of the earth and of life on it. (It may consequently be more accurate to talk about an "aesthetic-historic" interest instead of just an "aesthetic" interest, but for the sake of briefness I will stick to the second term.)[15]

My proposal implies that, from the environmental ethics perspective, we are more concerned about beautiful and spectacular organisms, like the snow leopard or the sequoia, than ugly and inconspicuous organisms. This implication is borne out by experience. The theory predicts that we are inclined to take a greater environmental interest in beautiful species which are unique, in the sense that there are no other species of the same family. As the beauty of these species is utterly different from the beauty of any other species, and these species can be the source of interesting information that cannot be obtained from elsewhere.

If this aesthetic construal of environmental concern is right, it is easy to explain why it does not cover the killing of individuals that belong to species with numerous members. Even if these individual members are quite spectacular – like members of the black-billed magpie (*pica pica*) – each individual contributes a drop to the beauty of the world at present and in the future, since the species could reproduce in adequate numbers without the participation of any particular individual. In contrast, if a spectacular species is rare, each member of it represents the highest intrinsic value from the perspective of environmental concern, since each of them adds comparatively much to the beauty of nature at present and in the future.

## 6. Environmental Concern instead of Environmental Ethics

This aesthetic value must be kept apart from the value that, according to conativism, an organism has in virtue of its possession of satisfiable desires. The aesthetic value is a value that the environment has *for beholders of natural beauty*, especially human beings. (Here, I cannot pursue the interesting question why we find nature so beautiful.[16]) It is a value an organism can have though it is never conscious and in possession of desires, for example, the sequoia. Therefore, the aesthetic approach to environmental ethics can easily explain why individuals could permissibly be treated harmfully, as means to the preservation of their species or other species. For this may promote the end of making the world more beautiful, and the aesthetic approach to environmental ethics can explain why this branch of ethics is straightforwardly consequentialist.

Earlier on I rejected anthropocentrism in the shape that only human welfare morally counts. But if my aesthetic approach to environmental ethics is correct, anthropocentrism returns to environmental ethics, for the appeal to aesthetics is anthropocentric. We have reason to preserve the natural environment because we find it beautiful and a rich source of information. The view that I am propounding should be distinguished from the view that beauty itself, independently of any beholder's appreciation, is a ground for moral status. It is tempting, but mistaken, to reason as follows. Things that are, for instance, beautiful have intrinsic value. If something has intrinsic value, it is intrinsically wrong, or wrong independently of effects on things external to it, to destroy it. But this is what it means to say that something has moral status or standing, that actions can be assessed as right or wrong, on the basis of their effects on this thing alone and independently of their effects on external things.

Michael Tooley is committed to something like this reasoning. For he writes that, "alongside the fact that things of a certain some type 'possess a right not to be destroyed,' the fact that 'things of that type are intrinsically valuable' is something that makes it 'intrinsically wrong to destroy' things of that type."[17] But whereas it is in itself objectionable to destroy something that has a right not to be destroyed, it need not be in itself objectionable to destroy something of intrinsic value because, as I have contended, this value may be a value for something else.

To say that something has a value, that is, intrinsic should be taken to mean only that its value does not derive from the value of anything external to it. This is however compatible with the intrinsic value being *relative*, being a value for some subject external to the thing with intrinsic value. For instance, the conativism I have put forward, roughly defines the notion of a thing T's being of value for a subject S in terms T's satisfying some desires of S's. If this is a desire for T for its own sake, T will have intrinsic value for S. This is a relativist interpretation of intrinsic value, in the sense that something's intrinsic value is *constituted* by its relation to something else (namely, the relation of fulfilling some desire). On this interpretation, the value of T is still intrinsic, since it is not derived from the *value* of anything external to T. The relativist interpretation is a claim about the *analysis* of value and, so, must not on pain of circularity employ the concept of value in its *analysans*. In contrast, a characterization of what distinguishes intrinsic from extrinsic value, is a distinction within the category of value which may legitimately treat the concept of value as primitive or undefined. (These distinctions are unfortunately obscured by G. E. Moore's famous definition of intrinsic value: "To say that a kind of value is 'intrinsic' means merely that the question whether a thing possesses it, and in what degree it possesses it, depends solely on the intrinsic nature of the thing in question."[18])

If we want to allow for the truth of the relativist interpretation of intrinsic value, we cannot say that, if T is intrinsically valuable, it must be possible for it to be intrinsically wrong – that is, wrong independently of effects on things external to T – to destroy T. For if the intrinsic value T has is a value for S, it is the effect on S – that it frustrates one of S's desires – that makes the destruction of T wrong. So, even if T has intrinsic value, the destruction of T will not be intrinsically wrong. Therefore, it does not follow that T has moral status in the sense that something has moral status, if and only if it can be wrong to damage it independently of the effects on other things.

According to conativism, it can be intrinsically wrong to destroy something only if it has (satisfiable) desires. For only then can it be a subject of (positive) value or valuation in the sense that something can have value for it. It then becomes possible that the destruction of this subject be intrinsically wrong. Incidentally, it might be misleading to speak of desiring something "for its own sake" when something is intrinsically desired. For this manner of speaking might mislead people to think that the object desired has a "sake" or an end that one desires to fulfill. And then we seem to have an instance of genuine ethical concern. But this is not so, unless to have a "sake" implies having desires. However, it should not be taken to do so in the expression "desiring something for its own sake."

A corollary of this argument is that it is a misnomer to talk about environmental *ethics*. I have advocated the view that a proper object of ethical concern is a subject for whom something can be of (intrinsic) value, and I have equated such subjects with subjects who are at some time equipped with desires. It is only with respect to such conative subjects that actions can be intrinsically wrong, or wrong independently of their effects upon things external to these subjects. So, strictly speaking, instead of environmental ethics we should speak of something like an environmental *concern*. Only a proper subset of the objects of this concern – those entities with desires – are objects of moral concern, though they may have intrinsic value even if this is not so. If we thought that something is a bearer of intrinsic value was sufficient for it to be an object of moral concern, we would have ethics of different artifacts, like artworks, since they can have intrinsic value as much as natural things.

It is universally recognized that environmental concern can stand in opposition to concern for (in short-term human welfare. If my interpretation of environmental concern is right, we must take seriously the possibility that concern for *animal* welfare can also conflict with environmental concern. We indulge in anthropocentrism, if we take it for granted that our aesthetic ideals for the environment must promote the welfare of the animals that figure in the environment, or the environmental concern is necessarily aligned with animal ethics.

Although the view of Buddhists and Schopenhauer is excessively pessimistic, it is agreed by most that the evolutionary struggle for survival and reproduction that rules life on earth leaves much to be desired from the point of view of morality. When we preserve the natural environment, we maintain this merciless struggle. This is comparable to preserving outmoded, say, feudal, forms of social life along with the old buildings that were the stage for these life forms. Most of us would agree that could be deeply immoral. Similarly, it could be deeply immoral to preserve the natural environment.

It would be so, if Buddhists and Schopenhauer are right in their equation of life and suffering. We would then be under a moral obligation to put an end to conscious life on earth. Few of us believe in this equation. But those of us who are concerned about preserving the natural environment must believe not only that for most sentient beings their natural existence is better than non-existence, but also that it is better than it would be on any (non-natural) arrangement that we could realistically implement. This is hard to believe. But if so, we face an uncomfortable conflict between concern for animal welfare and environmental concern. For we would be morally obliged to interfere with the beauty of a pristine natural environment to install this more moral arrangement. If this artificial arrangement of nature could be made as beautiful as the untouched natural environment, there would be even less to adduce against the implementation of it; there would be only historic considerations.

## NOTES

1. Bryan G. Norton, *Why Preserve Natural Variety?* (Princeton: Princeton University Press, 1987).

2. Peter Singer, *Practical Ethics* (Cambridge: Cambridge University Press, second edition, 1993), Ch. 3.

3. Ingmar Persson, *The Retreat of Reason: A Dilemma in the Philosophy of Life* (Oxford: Oxford University Press, 2005), Pt. II.

4. *Ibid.*, Ch. 4.

5. Michael Tooley, *Abortion and Infanticide* (Oxford: Clarendon Press, 1983), p. 67.

6. Paul W. Taylor, *Respect for Nature* (Princeton: Princeton University Press, 1986).

7. Holmes Rolston III, *Environmental Ethics: Duties to and Values in the Natural World* (Philadelphia: Temple University Press, 1988).

8. Mark H. Bernstein, *On Moral Considerability: An Essay on Who Morally Matters* (New York; Oxford: Oxford University Press, 1998), p. 91.

9. J. Baird Callicott, "The Search for an Environmental Ethic," *Matters of Life and Death: New Introductory Essays in Moral Philosophy*, ed. Tom Regan (New York; London: McGraw-Hill, third edition, 1993), p. 365.

10. *Ibid.*, p. 367.

11. *Ibid.*, pp. 375–376.

12. Immanuel Kant, *Grundlegung zur Metaphysik der Sitten*, p. 429. Herbert J. Paton, trans., *The Moral Law* (London: Hutchinson's University Library, 1948).

13. Alan Donagan, *The Theory of Morality* (Chicago: Chicago University Press, 1977), Ch. 5.

14. Arthur Schopenhauer, *Die Welt als Wille und Vorstellung*. E. F. J. Payne, trans., *The World as Will and Representation* (New York: Dover Publications, 1969). And Arthur Schopenhauer, *Über das Fundament der Moral*; and also. E. F. J. Payne, trans., *On the Basis of Morality* (Providence; Oxford: Berghahn Books, 1995).

15. Lily-Marlene Russow, "Why Do Species Matter?" *Animal Rights and Human Obligations*, eds. Tom Regan and Peter Singer (Englewood Cliffs, N.J.: Prentice Hall, second edition, 1989).

16. John Andrew Fisher, "Aesthetics," *A Companion to Environmental Philosophy*, ed. Dale Jamieson (Malden, Mass.: Blackwell, 2001).

17. Tooley, *Abortion and Infanticide*, p. 95.

18. George Edward Moore, "The Conception of Intrinsic Value," *Philosophical Studies* (London: Routledge & Kegan Paul, 1922), p. 260.

# Four

# EMPIRICAL ENVIRONMENTAL ETHICS

Y. S. Lo

## 1. Introduction

In popular terms, "moral relativism" refers to the view that no across-the-board moral standards exist: values and disvalues, rights and wrongs, are relative to different groups, and, in the extreme case, to individuals. The growing attractiveness of this position – especially to younger generations living in cosmopolitan societies – reflects the existence of wide and apparently irresolvable moral disagreements among people from different social, cultural, and religious backgrounds. Relativism and its underlying skepticism to universality can sometimes function as an antidote to fundamentalism, challenging an authority's imposition of supposedly absolute and universal standards on the unwilling others. But when adopted as an automatic default position, relativism runs the risk of moral isolation, blocking external critique and diminishing internal reflection. Just as global superpowers can use the rhetoric of self-evident universal values (such as "universal human rights" and "fair trade") for imperialistic advancements, regional powers can use the rhetoric of relative values (such as "Asian values" and religious values) to sustain repressive practices.

Opposed to popular relativism and skepticism, and responding to the phenomenon of evaluative divergence, is an optimistic belief embedded in western liberalism, shared by many contemporary intellectuals. This is the universalist belief that human nature is sufficiently uniform across cultures and societies so that people with incompatible ethical and political stances, if put under favorable conditions (such as Juergen Habermas's "ideal discourse situations" (Habermas, 1974 and 1990), John Rawls's "veil of ignorance" (Rawls, 1971, 1985 and 1993), and Michael Smith's "full rationality" (Smith, 1994)), will eventually converge in their beliefs and attitudes (or achieving what Anthony Giddens calls "dialogic democracy" (Giddens, 1994), or what Richard Rorty calls "social solidarity" among "liberal and tolerant fuzzies" (Rorty, 1989 and 1991)).

The present project makes no assumptions about whether the optimistic liberal view on the uniformity of human nature is true. Instead, it regards that as an empirical matter which cannot be settled by theorists *a priori*. It notes that no systematic and commonly accessible methods presently exist for answering moral questions in the first place. It is the lack of such methods that has led to the growth of skepticism and relativism in the face of continuing moral disputes.

## 2. Objectives

As an antidote to moral skepticism, imperialistic universalism, and defensive relativism, the project's first general objective is to provide a conceptual framework within which systematic empirical methods can be devised for determining (1) whether a putative value is a genuine value, and if so, whether it is universal or relative, and then (2) how competing values are to be prioritized.

The idea of using concrete empirical methods to answer abstract moral questions is central to the recently emerged interdisciplinary area of empirical ethics. From an empiricist perspective, the project argues that many moral questions about good and evil, right and wrong, are ultimately reducible to empirical questions about some human psychological capacities, tendencies, and habits. Hence inquiries into value and morality require us to go beyond pure reasoning and logical arguments, and step into the empirical world to observe and investigate those aspects of the human being. This can only be done via collaboration among the psychological and social sciences, the biological and neurological sciences, and history and philosophy. By identifying particular aspects of the human psyche, and explaining how they are constitutive of the ethical domain, the project will define new directions for interdisciplinary researches in empirical ethics.

Because empirical ethics takes scientific methodology so seriously makes it suited to the robust character and naturalistic tradition of analytic philosophy. Little interest in this new area of research, however, has been showed by moral philosophers in such traditions. The project's second general objective is to communicate information about, and the development of, empirical ethics to the wider philosophical community in Hong Kong and Australia. Both are places where I have previously conducted relevant research.

## 3. Working Hypotheses

The project is humanist. It takes ethics to be fundamentally about human beings (instead of about God or some super agent devoid of human needs and weaknesses). The first working hypothesis is that the purpose of ethics is to answer how human beings can co-exist in sustainably happy and flourishing ways, and to assist them doing so (cf. Hume, 1739-40 and 1751; MacIntyre, 1984; Williams, 1985 and 1995).

The project is also dispositionalist. Its second working hypothesis is that core moral concepts can be analyzed and defined in terms of people's dispositions to give affective responses under a set of favorable conditions (cf. Lewis, 1989; Johnston, 1989; Smith, 1994; Elliot, 1997). A useful analogy is a form of dispositional account of colors, which understands an object as having a particular color (for example, red) on the basis that normal people are disposed, under the favorable conditions for observing colors (for example, under normal day light), to experience the object as looking that color. Similarly, we can understand an object as having a particular moral property (for example, value) on the basis that normal people are disposed, under the favorable conditions for making moral evaluations, to have corresponding affective response (for example, approbation) towards the object.

The choice of the humanist hypothesis and the dispositionalist hypothesis will integrate well with the naturalistic and reductive approach of empirical ethics. It will also permit the project to contribute to the scholarship of David Hume, which is significant in its own right.

## 4. A Humanist Dispositional Theory of Value

Based on the second hypothesis just discussed, the project will develop a dispositional theory of value, which analyses the core moral concepts "value" and "disvalue" (and other interdefinable concepts such as "obligation," "right" and "wrong," "virtue" and "vice"). The schema of the theory is as follows:

T. X is (relatively/universally) valuable/disvaluable just if (some/all) human subjects are disposed, under favorable conditions {C}, to feel the sentiment of approbation/disapprobation towards X.

A parallel analysis of "obligation" will state:

> The pursuit/avoidance of X is (relatively/universally) obligatory just if (some/all) human subjects are disposed, under favorable conditions {C}, to feel the sentiment of approbation/disapprobation towards X.

Using T as a general schema, we can also formulate parallel analyses for other interdefinable moral concepts. But let us focus on T and look at it in more details.

First, different kinds of things can be valuable/disvaluable. Character traits constitute one kind. Valuable character traits are virtues; disvaluable character traits are vices. Actions and behaviors can also be valuable/disvaluable. The morally right ones are valuable; the morally wrong ones are disvaluable. Similar considerations apply to material and abstract objects, events, and states-of-affairs. As an analysis of the general notions "valuable" and "disvaluable," T covers all sorts of things to which these notions can be sensibly applied.

Next, to say "we are disposed under such-and-such condition to do such-and-such" is just an abbreviated way of asserting "if we are (or were, or had been) under such-and-such condition, then we will do (or would do, or would have done) such-and-such." A standard way of testing a statement like this is to meet the stated condition, and then check whether the expected result or something similar follows. So, given T, in order to find out whether X is a value, we need to find out whether we are disposed under conditions {C} to feel approbation towards X. In order to find out whether we are so disposed, we need to do two things. First, we try our best to meet the approximate conditions {C}. Second, while under those conditions, we see as honestly as possible what evaluative responses we have towards X. We may feel approbation/disapprobation towards it, or we may be indifferent. According to T, approbation/disapprobation under conditions {C} indicates value/disvalue, whereas indifference under such conditions indicates non-value.

Then, what exactly are the favorable conditions, which we must meet the approximate in order to make correct or close-to-correct moral evaluations? The project proposes the following three.

C1. The condition of being empathetically aware of relevant facts and relations, *vis-a-vis* the object under evaluation. (Object-awareness Condition)
C2. The condition of being empathetically aware of basic facts about human nature. (Human-nature-awareness Condition)
C3. The condition of non-egocentrically considering one's particular interests and relations *vis-a-vis* the object under evaluation. (Decentering Condition)

These conditions are not dissimilar to those that a competent judge or jury in the court room is expected to meet. The project argues that the choice of the above conditions as favorable for making moral evaluations is directly supported by the humanist hypothesis (section 3 above). Suppose the hypothesis is right in taking ethics as a system conducive to happy and flourishing human lives. Then moral evaluations (which are supposed to give us guidance on how to live) must reflect the basic facts of human nature (for example, their basic needs and desires, weaknesses and strengths) that set the perimeters within which human beings can live happy and flourishing lives. This explains C2's emphasis on the awareness of basic facts about human nature. Suppose the hypothesis is also right in taking ethics as advantageous for human co-existence. Then, moral evaluations (which are supposed to guide our interactions with each other) must be capable of being adopted by different human individuals despite their diverse temperaments, interests, and circumstances. One obvious way to facilitate the sharing of views and attitudes is to incorporate the operation of empathy as part of the proper communication process. Empathy is the ability of entering into another's personality. It has two components. The cognitive component involves accurately detecting another's feelings and perspectives. The affective component involves using imagination to experience those feelings vividly and take for oneself those perspectives that one detects in another. As a result, the greater the empathy people have for each other (both cognitively and affectively), the more likely they are to come to agreement. This explains C1 and C2's emphasis on empathic awareness. But a pre-condition for empathy is decentering – the ability to set aside temporarily one's particular attachments. Decentering must not be confused with self-denial. The second requires neglecting one's self-interests and relations. But the first requires considering and counting one's self-interests and relations, but in a non-egocentric manner – that is, count them as a third party would count them. This is what is meant by C3's requirement of non-egocentricity.

Now, given T's analysis of value in terms of people's dispositions to respond under the specified favorable conditions, we can see that a major source of evaluative error is that people make evaluations under less than favorable conditions. To use the analogy of color again. An inexperienced observer might mistakenly judge a red object to be brown because of dim lighting. Similarly, an inexperienced evaluator might fail to feel disapprobation towards a bad behavior and so judge it to be morally neutral. For this inexperienced evaluator might be evaluating it under less than favorable conditions. For instance, the evaluator might dislike, and so have low empathy for, the person at the receiving end of the behavior (that is, condition C3 not met). Just as not meeting the favorable conditions can cause error in our moral judgments, better satisfying those conditions can make our judgments more reliable. Due to new awareness of some basic facts of human nature (and so better satisfying condition C2), person might give up their originally negative moral judgment of some kind of behavior (for example, homosexuality). In short, the progress of our moral knowledge depends to a large extent on the progress we make on meeting those three favorable conditions. It is unrealistic to expect people to satisfy those conditions fully whenever they make moral evaluations. The main point of T, as we get better at meeting or approximating those conditions – by trial and error, accumulated knowledge and experience of human life, and continual practice of empathy – is that our sentiments of approbation and disapprobation will become better indicators of value and disvalue.

Note that T gives only a qualitative analysis of the concepts "value" and "disvalue," the function of which is to find out what values exist. But that is not enough for solving ethical problems about value. For even if we could identify all the values in the world, by testing people with methods based on T, we would still have the problem of how to prioritize conflicting values. For example, a pristine wilderness area free from human interference might have a value in itself that should be preserved; but the opening of the area to travelers, researchers, or developers might bring into being other values, such as aesthetic pleasure derived from visiting the area, improvement of human health because of the discovery of ingredients in the area for new medicines, or an increased standard of living for the local community due to the development of new infrastructure in the area. So, in order to solve the problem of conflicting values, which is especially relevant to social policy-making, a quantitative analysis of value is also needed. Based on the two working hypotheses discussed before, the project proposes the following general schema for a quantitative analysis of value.

T*. X is (relatively/universally) more/less valuable/disvaluable than Y just if (some/all) human subjects are disposed, under favorable conditions {C*}, to feel more/less of the sentiment of approbation/disapprobation towards X than towards Y.

Since prioritizing different values is quite a different task from identifying them in the first place, the project will look into the possibility that testing people under condition T* may need a different set of "favorable conditions" from those associated with examining people under T. After both, possibly over-lapping, sets of favorable conditions are fully specified, the complete versions of T and T* will then set the conceptual framework within which systematic and empirical methods can be developed for finding out (1) the values of a population of any size, and (2) the priorities over those values.

## 5. Applying the Theory

Building on my previous work on environmental ethics (Lo, 1999 and 2001c) and current work on bioethics, the project seeks to use the dispositional theory just described to analyze and assess value-conflicts in ethical debates concerning (1) conservation and land management, and (2) the application and development of biotechnologies.

We have seen that the proposed dispositional theory of value reduces the study of moral value and obligation to an empirical study of people – more precisely, their dispositions to give some kind of affective responses under the three humanist favorable conditions, C1, C2, and C3. In order to apply the dispositional theory to find out the conflicting values involved in a moral debate, the first thing we need to do is to assist the participants in the debate to put themselves under those three humanist favorable conditions. These conditions, as we have seen, are conducive to honest and empathic communication, which will in turn increase the chance of convergence in beliefs and attitudes. The process of using the theory to test the values of people from diverse backgrounds on divisive issues will itself facilitate (though not guarantee) conflict resolution. And its application targeted populations or groups facing particular social or ethical problems should promote community-based and ethically informed policy-making. The project argues that active application of the proposed dispositional theory to people in real life situations of social, political, and moral conflicts will assist the advancement of modern liberal democracy towards its optimal forms, which can accommodate differences and yet retain standards – both within humanist boundaries.

## 6. Contemporary Metaethics

In the course of developing the dispositional theory of value, the project will also re-conceptualize and coherently answer a series of long-standing meta-ethical problems. In particular, the project defends the following four positions that are logical consequences of the dispositional theory. (1) It defends the view that value/disvalue is ontologically objective but conceptually subjective – that is, value/disvalue can exist in the absence of the evaluating subject's positive/negative response, although the concept of value/ disvalue cannot be analyzed or defined without referencc to the evaluating subject. (2) It argues that immorality is a particular kind of emotional – instead of a rational – deficiency. While some level of rationality is necessary for moral behavior (as it is necessary for many other kinds of behavior), no amount of rationality is sufficient for an ethical or good human life. (3) It argues that *a priori* reasoning and logical arguments cannot rule out the empirical possibilities of moral relativism (that is, the view that there exist no universal values that are morally binding for all human beings), and that the truth/falsity of any claims about universal and relative values can only be determined by empirical tests based on the proposed dispositional theory. (4) It argues that the only unifying denominator, if any, in terms of which different values can be weighed against each other, is an emotion (or a range of emotions) that is quantifiable and can be experienced by human beings under the favorable conditions for assessing those different values. Whether there exists such an emotion (or range of emotions), is an empirical matter of fact to be investigated by the psychological and neurobiological sciences.

The combination of these positions represents a novel stance in meta-ethics, and offers a useful way to recast, reconnect, and resolve classic debates between objectivists and subjectivists ((1) above), between rationalists and sentimentalists ((2) above), between universalists and relativists ((3) above), and between monists and pluralists ((4) above). Given the limited space here, I will elaborate a little more only on (1) and (3).

### A. Objectivism and Subjectivism

Give schema T, value is conceptually subjective in that it analyzes the concept "value" in terms of concepts of some features of subjects, and so makes the former conceptually dependent on the latter. But value is also objective in some senses under T.

In the *Routledge Encyclopedia of Philosophy*, objectivism of value is described as the view that "value is not a category distinctively attuned to human concern but exists independently of human interests."(Wallace 1998). In this sense, Humean value is objective. Given T, to say that X is valuable is equivalent to asserting the conditional statement that if people are, or were, or had been, in the favorable conditions C1, C2, and C3, then they will feel, or would feel, or would have felt, the sentiment of approbation towards X. This conditional assertion can be true even if no one has the sentiment of approbation towards X. For it might be the case that no one has yet met all the favorable conditions for giving reliable sentiments, but if they had met those conditions they would have approved of X. As lack of approval from people towards X does not imply that X is not valuable, similarly, T allows the possibility that people might incorrectly approve of something that is not genuinely valuable because they are not evaluating under the favorable conditions. For T allows that value can exist in the absence of subjects' positive evaluative response and that things can be valuable without being valued. This means that value is ontologically objective. Since T reduces facts about values to objective facts about the psychological dispositions of subjects, it is a form of cognitivism that makes a place for truths and mistakes about values. It explains how such mistakes come about and tells us how to improve.

In short, if T is correct, then value is both objective and subjective, but in ways compatible with, and complementary to, each other, which is, in my view, an advantage of the theory.

### B. Universalism, Relativism, Nihilism

Under schema T, all values are relative. Because universal values are just one particular kind of relative values – namely, values relative to all people. T implicitly allows metaethic relativism, the view that non-universal relative values are nonetheless as genuine values as universal ones. But whether a putative value is universal (or for that matter, whether it is a genuine value) remains an empirical question.

Now, unless moral universalism (the view that some values are universal) is held as a closed theory, we must not rule out the possibility of a genuine other, someone who has genuinely different evaluative dispositions from ours – that is, different moral sentiments even under the same favorable conditions. The possibility of a genuine other sets a limit to the universalist optimism. As no *a priori* reason exists for thinking that all people must invariably share the same set of evaluative dispositions, it would be unreasonable to insist that whenever someone evaluates differently from how we would evaluate, she must have failed to satisfy the ideal conditions. It is not true that people must belong to each other come what may! The possibility of evaluative diversity under favorable conditions is just as real as the possibility of evaluative convergence under those conditions.

Despite its allowance for moral relativism (the view that some values are non-universal), schema T in no way suggests that we should not act according to our values or take measures to encourage others to adopt them even if they are not universal. People's evaluative dispositions are evolutionary and cultural products, and the products of personal history. They are not fixed absolutes but malleable to some extent. If T is right in understanding values as fundamentally anchored by people's evaluative dispositions, then values can be created and relative values can become more universal, to the extent that people can cultivate, negotiate about, and converge in, their evaluative dispositions.

## 7. Hume Scholarship

The project is Humean as much as it is humanist. It aims to contribute to contemporary scholarship on the 18th century empiricist philosopher David Hume, whose work has been a major influence on the development of modern analytic and empiricist philosophy. Building on my previous work on Hume (Lo, 2001a, 2001b, and 2006a), the project seeks to update his work on moral philosophy through its engagement with recent debates in metaethics and the new findings of empirical ethics.

In the course of developing the dispositional theory of value, the project draws upon theoretical resources from Hume, and advances a new controversial interpretation of his work on virtue and vice, moral motivation, and the purpose of morality. Against a well-known set of orthodox readings of Hume since the 1960s (for example, Hare, 1963; Foot, 1963; Flew, 1969; Harrison, 1976; Mackie, 1980; Ayer, 1980; Brown, 1988; Snare, 1991; Bricke, 1996; Wallace, 1998; and Baillie, 2000), the project will argue for the following four interpretations of Hume. (1) Hume himself does not endorse the so-called "Hume's Law" – the thesis that no evaluative "ought" can be derived validly from any empirical "is." (2) He is a cognitivist and a descriptivist – that is, he thinks that moral judgments (unlike expressions of emotion such as "boo," "hurrah") are capable of being true or false, and that they describe some matter of facts about the world. (3) He is neither an internalist nor an externalist – that is, he does not think that all moral judgments by themselves can motivate people who make those judgments to behave accordingly, nor does he think that no moral judgment can do that. For him, only a special class of moral judgments is intrinsically motivating without any external help. (4) Finally, he upholds a version of the dispositional theory of value as previously described.

This scholarly aspect of the project is intended to confirm Hume's pioneering empiricist and quasi-scientific approach to ethics, and argue that it provides a historical precedent for the recently emerged interdisciplinary researches in empirical ethics.

## 8. Approach and Methodology

The project argues for, and employs, a general model for effective ethical inquiry, which assigns a particular role to empirical ethics in relation to existing areas of moral philosophy. This general model distinguishes four main areas of ethical inquiry. They are: (1) applied ethics, which seeks to come up with particular moral judgments about concrete moral issues (for example, abortion in the case of deformed fetus); (2) moral theory (or normative ethics), which seeks to come up with general moral principles about what kinds of things are valuable/disvaluable, or what kinds of behaviors are right/wrong; (3) empiricist metaethics, which seeks to analyze core moral concepts in empirical terms, and so reduces moral properties to empirical features of the world; and (4) empirical ethics, which seek to verify particular moral judgments and general principles, by empirically and scientifically identifying concrete or abstract objects that fit the empiricist metaethical analyses. The diagram below shows the interactions among the four areas of ethical inquiry:

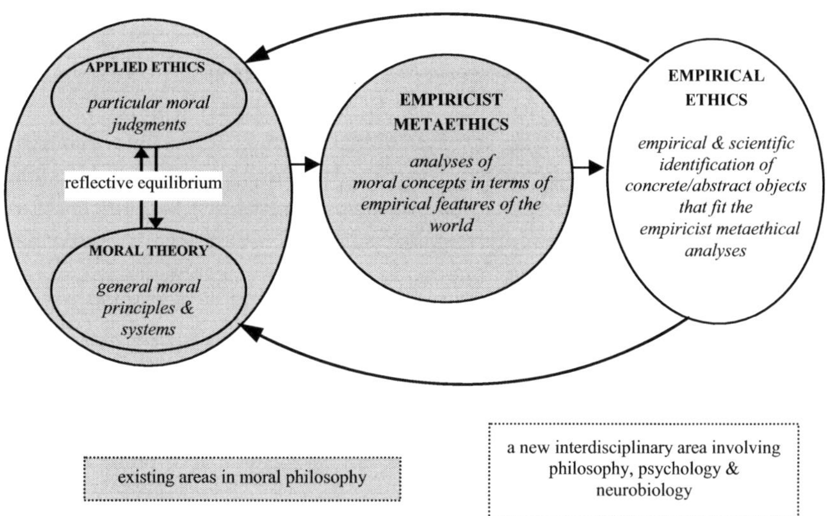


Under this model, effective ethical inquiry begins with some debates on particular moral issues. Such debates usually contain a mixture of (1) particular moral judgments in area of applied ethics (for example, in the abortion debate, the conservative judgment "abortion is wrong even if the fetus is severely deformed") and (2) general moral principles used in building systematic moral theories (such as "human life is valuable," "we should always act in such a way that maximizes values"). The ideal state of this mixture is the so-called "reflective equilibrium," where one's particular moral judgments are maximally consistent with one's general moral principles (cf. Rawls, 1971).

Then, based on the moral language(s) that we use to carry out the moral debates in the first place, we proceed to (3) empiricist metaethics, where we examine the core moral concepts (such as "value" and "obligation") used in those debates and other types of ethical discourse including those in fictions and films. In particular, we analyze the core moral concepts in empirical terms, and so reduce the corresponding moral properties to some empirical features of the world. For instance, the dispositional theory of value posited in the present project reduces the moral property of value to some empirical features of human beings. It is crucial to recognize that a good conceptual analysis needs to capture "folk platitudes" about the concept being analyzed (Jackson, 1998). These platitudes about moral concepts can be found in real ethical debates in the society, and also in debates among applied ethicists in more academic venues. But folk platitudes about apparently the same concept might be diverse. So the study of people's real reactions in ethical debate is essential for deciding which set of platitudes are more central than others. In cases where conflicting platitudes appear to be equally central, the concept in question may be vague or ambiguous, so requiring further analysis (cf. Brennan, 2004).

Next, building on the empiricist metaethical analyses of moral concepts, we proceed to (4) empirical ethics, where we employ empirical and scientific means to identify which objects fit or do not fit the metaethical analyses. We find out whether an object (be it concrete, say, a forest, or abstract, say, a form of life) has some moral properties (for example, the property of being valuable) by finding out whether it has their empirical equivalents (that is, whether it is such that people are disposed under favorable conditions to approve of). To use an anti-abortion and anti-euthanasia premise as an illustration: the analysis given by the project tells us that the only way to assess the moral principle "human life is always valuable no matter what" is to put people under conditions C1, C2, and C3, which are favorable for making evaluations, and then to check their subsequent responses towards different forms of human life. Human life is always valuable no matter what if and only if people are disposed under those favorable conditions always to value human life no matter what.

In term of the methodology for theory application, a useful comparison is Rawls's dispositional analysis of distributive justice. It says that "just" or "fair" principles of distribution of social goods are exactly those principles that people are disposed, under favorable conditions, to choose. The favorable conditions, he argues, are the conditions of being "behind a veil of ignorance" – that is, to lack information about one's particular characteristics, attachments, and position in society, and so be unable to make choices that are calculated solely to advance one's own self-interest. In the early 1990s, systematic empirical work in social and political sciences was carried out in the United States, Canada, Poland, and South Korea (Bond and Park, 1991; Frohlich and Oppenheimer, 1992) to test the Rawlsian analysis of justice. This involved using social psychological experiments with focus groups to locate people's preference behind a simulated "veil of ignorance." Similar empirical studies have been carried out recently in Beijing, Taipei, and Hong Kong (Chan, 2004; Lai, *et al*, 2003).

Consider the following parallels between Rawls's analysis and the one proposed here. In order to find out what principles (or objects) are just (or valuable), we need firstly to put people behind the "veil of ignorance" (or under "favorable conditions" C1, C2, and C3), and secondly see what principles (or objects) they choose (or approve of). Although, the Rawlsian veil of ignorance is quite different from the favorable conditions specified by the present project, the research methods and techniques employed in previous applications of Rawls's analysis of justice can be borrowed and modified for the applications of the project's analysis of value. Comparative researches into people's evaluative dispositions under those favorable conditions promise to inform the debate between universalism and relativism, and answer the crucial questions about whether values are relative to social and cultural backgrounds, and the scope of any universal values. In addition, the project's analysis of value will supplement recent interdisciplinary research into the neurobiological mechanisms of moral cognition and evaluative responses (see, for example, Greene *et al*, 2001; Montague *et al*, 2002; Casebeer, 2003, Casebeer and Churchland, 2003). By specifying the conditions favorable for making sound moral judgments, the project will advance the design of neurobiological experiments so as to study people's values proper, not merely their initial, unconsidered evaluative attitudes.

Equipped with the findings in empirical ethics about people's evaluative dispositions (whether they are delivered by social-psychological sciences or neuro-psychological sciences), we can go back to assess the moral debates that originally started this process of ethical inquiry. The empirical evidence from using the metaethical analyses to test real people's evaluative dispositions might confirm the applied-ethical judgments and the general moral-theoretical principles that we have in the beginning of the debates. Or the empirical evidence might disconfirm them and require their revision. For example, if findings from empirical ethics suggests that people are disposed, under the favorable conditions C1, C2, and C3, to value only some forms of human life but not others, then we will need to replace our initial moral principle "human life is always valuable no matter what" by a more restricted claim like "only some forms of human life are valuable." And depending on whether the deformed fetus in question is likely to have some valuable form of human life, we will also need to revise our initial applied-ethical judgment about whether abortion in the case of deformed fetuses is wrong.

Under the present model of effective ethical inquiry, the interchange among (1) applied ethics, (2) general moral theory (or normative ethics), (3) empiricist metaethics, and (4) the newly emerged empirical ethics, is an ongoing and expanding process of positive feed backs, constantly aiming at a coherent viewpoint for connecting and assessing a diverse range of social ethical problems.

By advancing empirical ethics in different communities, and providing a framework for developing systematic methods of resolving moral disputes, the project aims to ameliorate the problem of moral skepticism characteristic of contemporary society. As previously explained, the applications of the project's humanist framework to find out people's evaluative dispositions (for example, regarding conservation management and the use of biotechnologies) will raise public participation, and awareness of the importance of ethical considerations, in the process of policy-making. Such applications will also increase the likelihood of people's convergence in beliefs and attitudes, and assist the progress of liberal democracy.

## 9. From Environmental Virtues to Environment Values: An Open Question

Hume has argued that in the case of the virtue of justice, relevant moral sentiments can be cultivated and solidified via the communicative mechanism of sympathy (see Lo, 2001a). If that is the case, then it is plausible that those sentiments can also be internalized via the same mechanism to such an extent that they will survive the reflections under the conditions C1, C2, and C3, which are favorable for having reliable moral sentiments. Our moral psychology is malleable regarding not only our surface evaluative attitudes, but also our latent evaluative dispositions – though it is reasonable to expect the attitudes to be more flexible than the dispositions. In short, if schema T is correct in grounding values on people's evaluative dispositions, then values are to some extent to be made.

Hume's account of justice as an artificial virtue is a useful model for practical environmental ethics. Just as Hume conjectures that justice is a virtue partly because we have made it a virtue out of necessity and keep it so by continually participating in its conventions with both our moral sentiments and behaviors, one may reasonably argue that we need to make some environment-friendly character traits virtuous so to ensure the continuation of human prosperity. Examples of prospective "environmental-virtues" are modesty and thoughtfulness regarding the Earth's natural environment, which are expressed in activities such as recycling and reducing the consumption of petrol and other products that cause air and water pollutions. Like the virtue of justice (the scope of which is defined by its social conventions and the initial motive for participating in which is people's regard to self-interest), the emergence of some environment-friendly character traits as social virtues must be supported by conventions; and the initial motive in society for environment-friendly behaviors is likely to be egocentric and anthropocentric. On the necessity of justice as a virtue, Hume writes:

> [H]owever single acts of justice may be contrary [. . . to] private interest, "tis certain, that the whole plan or scheme is highly conducive, or indeed absolutely requisite, both to the support of society, and the well-being of every individual." Tis impossible to separate the good from the ill. [. . . E]very individual person must find himself a gainer, on balancing the account; since, without justice, society must immediately dissolve, and every one must fall into that savage and solitary condition, which is infinitely worse than the worst situation that can possibly be suppos'd in society. (Hume, 1739-1740, 3.2.2.22 (cited by paragraph, section, part, and book))

Analogously, the necessity of environmentally-virtuous conventions can be put as follows:

> However much a single exercise of environmental virtues, such as modesty and thoughtfulness, may be contrary personal interest and short-term human interests, it is certain that the whole plan or scheme is highly conducive, and indeed absolutely requisite, to the support of the Earth's natural environment, the human civilization, the survival of the Homo Sapiens species, and the well-being of every human individual. It is impossible to separate the good from the ill. Every individual person must find themselves a gainer, on balancing the account; since, without the rules and conventions of those environmental-virtues, the natural environment and human society must soon dissolve, and every one must soon fall into that savage and impoverished condition, which is infinitely worse than the worst situation that can possibly be supposed in an environmentally-virtuous society.

Arguably, the egocentric and anthropocentric justifications for environment-friendly behaviors are only the beginning of any durable environmentally-ethical practice. As Hume points out, the continual practice of the conventions of justice in larger societies requires the development and internalization of a non-egocentric sense of justice as a virtue, which motivates just behaviors even when that happens to be against self-interest. Similarly, if environment-friendly conventions are to survive in modern cosmopolitan societies where community ties and sanctions are weak and the temptation of environment-damaging life styles is constant, then it is likely that people will need to develop not only a non-egocentric sense, but also a non-anthropocentric sense of virtue in following those conventions and vice in violating them. This non-anthropocentric sense of different environmental virtues and vices is in effect a set of internalized dispositions to experience different moral sentiments, under the favorable conditions C1, C2, and C3, towards different objects and behaviors. For instance, to have a non-anthropocentric sense of environmental-modesty as a virtue, and environmental-wastefulness as a vice, is, among other things, to be disposed to feel proud for taking the trouble to switch to solar energy, or for reducing the use of private vehicles by taking public transport instead, or to feel guilt and shame for forgetting to turn off the air-conditioning before leaving home or office, or for making unnecessary photocopies – even if the effects of such activities on oneself or other human beings are negligible. For another example, to have a non-anthropocentric sense of respect for nature as a virtue is to be disposed to feel the moral sentiments of disapprobation towards activities that endanger other species or destroy their natural habitats, and approbation towards activities that preserve those things – even if their destruction will cause little inconvenience to the human species.

The different moral sentiments that the environmentally-virtuous person is disposed under the Humean conditions to feel will then motivate her to behave virtuously towards the environment. Given schema T, because the environmentally-virtuous person is disposed under the favorable conditions to non-instrumentally approve of the preservation of the natural environment and its nonhuman inhabitants, in effect, constitutes the moral fact that these natural nonhuman entities are intrinsically valuable. The cultivation and internalization of a non-anthropocentric sense of environmental virtues not only produce environment-friendly behaviors but also create intrinsic values in the natural environment. Given T, (relative/universal) intrinsic values exist in nature if and only if some things in nature are such that (some/all) people are disposed, under the favorable conditions, to experience the moral sentiments of approbation towards them. Likewise, people can create (relative/universal) values in nature if and only if (some/all) people can successfully cultivate and internalize the corresponding psychological dispositions. But can people create whatever value they will? Regarding the development of the moral sentiments and dispositions that make justice a valuable character trait, Hume writes:

> [I]f nature did not aid us [. . .] twou'd be in vain for politicians to talk of honourable or dishonourable, praiseworthy or blameable. These words wou'd be perfectly unintelligible, and wou'd no more have any idea annex'd to them, than if they were of a tongue perfectly unknown to us. The utmost politicians can perform, is, to extend the natural sentiments beyond their original bounds; but still nature must furnish the materials, and give us some notion of moral distinctions. (Hume, 1739-1740, 3.2.2.25)

The same point applies to environmental ethics – especially the non-anthropocentric theories that attribute intrinsic value to the nonhuman nature. If valuable things are those towards which we are disposed to feel the pleasing sentiments of approbation, but if we are not already bestowed with some natural tendency to appreciate the natural world and/or its nonhuman inhabitants as ends-in-themselves, then it will be in vain for philosophers to argue that those things have intrinsic value and moral standing – these notions will be perfectly unintelligible and will not have any sense annexed to them. So, if schema T is correct in reducing values to human evaluative dispositions, then the most that non-anthropocentric environmental philosophers can do is to be informed and realistic about human evaluative psychology, and try to excite the relevant moral sentiments from us towards the objects of environmental concern by catching our imagination and sympathetic understanding with their philosophical arguments, and hope that those sentiments will endure our reflections under the conditions favorable for making moral evaluations. But after all, this can happen only if there are already some original materials of our moral psychology that underline our dispositions (if any) to appreciate, respect, and care for those things, and give us some notion of their moral significance. Whether, and to what extent, there are any such original ingredients of the human psychology, however, are more properly seen as empirical (and therefore open) questions, which are not single-handedly answerable by philosophers *a priori*.

## WORKS CITED

Ayer, A. Jules. (1980) *Hume.* Oxford: Oxford University Press.

Baillie, James. (2000) *Routledge Philosophy Guidebook to Hume on Morality*. London: Routledge.

Beauchamp, Tom L. (ed.) (1998) *David Hume: An Enquiry Concerning the Principles of Morals*. Oxford: Oxford University Press.

Bond, Doug and Jong-Chul Park. (1991) "An Empirical Test of Rawls's Theory of Justice: A Second Approach, in Korea and the United States," *Simulation & Gaming*, 22, pp. 443–462.

Brennan, Andrew. (2004) "Biodiversity and Agricultural Landscapes: Can the Wicked Policy Problems Be Solved?" *Pacific Conservation Biology*, 10, pp. 124–144.

Bricke, John. (1996) *Mind and Morality: An Examination of Hume's Moral Psychology*. Oxford: Clarendon Press.

Brown, Charlotte. (1988) "Is Hume an Internalist?" *Journal of the History of Philosophy*, 2, pp. 69–87.

Casebeer, William D. (2003) *Natural Ethical Facts: Evolution, Connectionism, and Moral Cognition*. Cambridge, Mass.: MIT Press.

________ and Patricia S. Churchland (2003) "The Neural Mechanisms of Moral Cognition: A Multiple-Aspect Approach to Moral Judgment and Decision-Making," *Biology and Philosophy*, 18, pp. 169–194.

Chan, Ho-Mun. (2004) "The Ethics and Care and Political Practices in Hong Kong." In: Beng-Huat Chua (ed.), *Communitarian Politics in Asia*, pp. 102–121. London: RoutledgeCurzon.

Elliot, Robert. (1997) *Faking Nature: The Ethics of Environmental Restoration*. London: Routledge.

Flew, Antony. (1969) "On the Interpretation of Hume." In: William Donald Hudson (ed.), *The Is-Ought Question: A Collection of Papers on the Central Problem in Moral Philosophy*, pp. 64–69. London: Macmillan.

Foot, Philippa. (1963) "Hume on Moral Judgment." In: David Francis Pears (ed.), *David Hume: A Symposium*, pp. 74–80. London: Macmillan.

Frohlich, Norman and Joe A. Oppenheimer (1992) *Choosing Justice: An Experimental Approach to Ethical Theory.* Berkeley; Oxford: University of California Press.

Giddens, Anthony. (1994) *Beyond Left and Right: The Future of Radical Politics*. Cambridge: Polity Press.

Greene, Joshua D., *et al.* (2001) "An fMRI Investigation of Emotional Engagement in Moral Judgment," *Science*, 293, pp. 2105–2108.

Habermas, Jürgen. (1974) *The Theory of Communicative Action*, trans. Thomas McCarthy. Vol. 1. Boston: Beacon Press.

________. (1990) *Moral Consciousness and Communicative Action*, trans. Christian Lenhardt and Shierry W. Nicholsen. Cambridge, Mass.: MIT Press.

Hare, Richard Mervyn. (1963) *Freedom and Reason.* Oxford: Oxford University Press.

Harrison, Jonathan. (1976) *Hume's Moral Epistemology*. Oxford: Clarendon Press.

Hume, David. (1739–40) *A Treatise of Human Nature*. Edited by David Fate Norton and Mary. J. Norton, Oxford University Press, 2000.

________. (1751) *An Enquiry Concerning the Principles of Morals*. Edited by Tom L. Beauchamp, Oxford University Press, 1998.

Jackson, Frank. (1998) *From Metaphysics to Ethics: A Defence of Conceptual Analysis*. Oxford: Clarendon Press.

Johnston, Mark. (1989) "Dispositional Theories of Value," *Proceedings of the Aristotelian Society*, Supp. Vol., 63, pp. 139–174.

Lai, Julian C., *et al.* (2003) "Dispositions toward Environmental Hazards in Hong Kong Chinese," *Journal of Environmental Psychology*, 23, pp. 369–384.

Lewis, David K. (1989) "Dispositional Theories of Value," *Proceedings of the Aristotelian Society*, Supp. Vol., 63, pp. 113–137.

Lo, Yeuk-Sze (1999) "Natural and Artifactual: Restored Nature as Subject," *Environmental Ethics*, 21, pp. 247–266.

________. (2001a) "Non-Humean Holism, Un-Humean Holism," *Environmental Values*, 10, pp. 113–123.

________. (2001b) "A Humean Argument for the Land Ethic?" *Environmental Values*, 11, pp. 523–539.

________. (2001c) "The Land Ethic and Callicott's Ethical System (1980-2001)," *Inquiry*, 44, pp. 331–358.

________. (2006a) "Making and Finding Values in Nature: From a Human Point of View," *Inquiry*, 46, pp. 123–147.

MacIntyre, Alasdair C. (1984) *After Virtue: A Study in Moral Theory*, second edition. Notre Dame, Ind.: University of Notre Dame Press.

Mackie, John L. (1980) *Hume's Moral Theory*. London: Routledge & Kegan Paul.

Montague, P. Read, *et al.* (2002) "Hyperscanning: Simultaneous fMRI During Linked Social Interactions," *Neuroimage*, 16, pp. 1159–1164.

Norton, David F. and Mary J. Norton (eds.) (2000) *David Hume, A Treatise of Human Nature.* Oxford: Oxford University Press.

Rawls, John. (1971) *A Theory of Justice.* Cambridge, Mass.: Belknap Press of Harvard University Press.

________. (1985) "Justice As Fairness: Political Not Metaphysical," *Philosophy and Public Affairs*, 14, pp. 223–251.

________. (1993) *Political Liberalism.* New York: Columbia University Press.

Rorty, Richard. (1989) *Contingency, Irony, and Solidarity.* Cambridge: Cambridge University Press.

________. (1991) *Objectivity, Relativism, and Truth.* Cambridge: Cambridge University Press.

Smith, Michael. (1994) *The Moral Problem.* Oxford: Blackwell.

Snare, Francis. (1991) *Morals, Motivation and Convention: Hume's Influential Doctrines*. Cambridge: Cambridge University Press.

Wallace, R. Jay. (1998) "Moral Motivation." In: Edward Craig (ed.), *The Routledge Encyclopedia of Philosophy*. London: Routledge.

Williams, Bernard. (1985) *Ethics and the Limits of Philosophy*. London: Fontana Press.

________. (1995) *Making Sense of Humanity and Other Philosophical Papers, 1982-1993*. Cambridge: Cambridge University Press.

# Five

# PERILS AND DANGERS: CLIMATE CHANGE AND THEOLOGICAL ETHICS

Michael S. Northcott

Lighten our darkness Lord we pray
And by thy great mercy
Defend us from all perils and dangers of this night
For the love of thy only Son
Our Savior Jesus Christ.[1]

## 1. Biogeochemistry and the Biblical Account of Moral and Spiritual Pollution

Scientists at the World Health Organization estimate that in the year 2000 around 160,000 people in the South were dying every year as a consequence of extreme weather events such as prolonged anticyclones and consequent flooding in South Asia, and extreme heat and drought in sub-Saharan African and the plains of India. Such events are becoming more frequent as a consequence of global warming and the WHO expects that the numbers of those dying from extreme weather events will double by 2020.[2] The Intergovernmental Panel on Climate Change (IPCC) suggest that up to three billion people, mostly in the tropics and the Southern hemisphere, will be displaced or suffer serious threats to their safety and security if greenhouse gas emissions continue to rise unchecked in the twenty-first century.[3] On even the most optimistic scenarios of the IPCC, established by the United Nations in 1989 in response to the threat of climate change, island nations including Kiribati, the Maldives, Mauritius, and the Seychelles, which rise no more than two meters above sea level, will cease to exist, and long before their land is inundated their water sources will have been turned to salt by the rising tides, rendering them uninhabitable.[4]

Although uncertainty remains about its precise effects climatologists are virtually unanimous that anthropogenic climate change is already a reality and that its effects will grow in intensity as the present century proceeds. However, consumers and corporations in the industrial North, who produce more than eighty percent of greenhouse gases, so far show little evidence that they collectively view climate change as a moral priority. While 187 nations have signed the Kyoto Protocol, the Protocol commits the heaviest polluters to reductions of only two percent on average in outputs of greenhouse gases between 2005 and 2012, reductions so small that they will only marginally reduce rising greenhouse gas production. Despite the catastrophic forecasts of the dangers that a warming world poses to human being and other species, the consumer society of the North, with its roots still firmly set in the colonial extractive economy, continues in its energy hungry path, and the peoples of the South continue to bear the cost.

Those who study the complex interactions between atmosphere, land, oceans, and species have founded a new discipline in the last twenty years, which they call biogeochemistry.[5] Biogeochemists in their description of climate change are at one with the Biblical story for they have found that ocean currents, the salinity and temperature of the oceans, and the quantity of plankton and other life forms they sustains are among the most critical determinants of the earth's climate. Ice core records show that at those times when the sea encroached over far more of the earth's surface than it does now, the earth did not provide the kind of stable climate that humanity has known for the past 15,000 years in the geological era known as the Holocene.[6] But before the Holocene, when mammoths roamed the earth alongside a quite small number of neolithic human beings, abrupt climate extremes of heat and cold were commonplace. Once the planet settled into a pattern of more gradual heating and cooling in response to movements in the orbit of the earth around the sun, the climate of the earth became more stable and it was possible for human settlement, and the human species, to gradually expand to the present state of human planetary dominance.

According to the creation myth of *The Book of Genesis*, the dominion of humanity, and the peace and order of creation which makes the expanding settlements of humanity possible, are a result of the providential ordering of God in calming and restraining the violent forces of the ocean, over which the Spirit of God is said to have first brooded, and so producing a stable climate and a fruitful earth. Contemporary biogeochemistry also shows that the ocean is the primary arbiter of the earth's climate. The ocean covers two thirds of the earth's surface. It is the principal heat store and cold store of the planet. Its currents are the primary means by which heat and cold are moved around from the tropics to the poles, warming some shores, cooling others. The reason that anthropogenic climate change is beginning to kick in is because of delayed reactions in the oceans to the build up in the last three hundred years of human pollution in the earth's atmosphere, reactions which are showing up in changing ocean temperatures and changing thermohaline currents, such as the Gulf Stream, which move between the warm tropics and the cool poles.

In *The Book of Genesis*, the covenant which God made with Noah and his descendants after the Flood promised that the earth would not again be threatened by the bursting forth of the chaotic waters. It was a "cosmic covenant" which included not only human beings but "living things of every kind."[7] As Robert Murray suggests, the idea of confrontation between the chaotic and elemental powers of ocean and climate and the ordering power of God is central to the Old Testament concept of covenant.[8] In the New

Testament water remains a powerful symbol of the ordering power of divine creation: when Christ stills the stormy waters of the Sea of Galilee the disciples ask "what sort of man is this, that even winds and sea obey him?" (Matthew 8:27).[9] Water also symbolizes the "new creation" which is the redeemed community of the new people of God. The baptized are loosed from the disorderly forces of the world, the flesh and the devil, and bound to Christ, when they pass through the waters of baptism, and the boundary between those who belong and those who do not belong to the body of Christ is defined by the holy waters of baptism, just as the ancient Hebrews reached the promised land by passing through the waters of the Red Sea.

The waters of baptism are the form in the church's life of the original cosmic covenant established between God and creation, a covenant which involved an intrinsic relationship between the human good, human sin, and the fruitfulness and order of the earth. The cosmic description of the covenant, like the Genesis stories of Adam and Eve, Cain and Able, train Jews and Christians to see that human beings are morally and biogeochemically implicated in the fate of the earth. Hence after Moses pollution is seen as a consequence of the failure to follow divine law: as Isaiah says, "the earth lies polluted under its inhabitants; for they have transgressed the laws, violated the statutes, broken the everlasting covenant" (Isaiah 24:5). Pollution here in biogeochemical terms indicates the salination of the soil, a consequence as we now know of excessive use of ground water, and of intensive agricultural practices on the deforested slopes and plains of ancient Mesopotamia which turned the "fertile crescent" into the near desert that covers much of the region today.

The region of Mesopotamia was the first where human beings developed agricultural civilization and it was this agricultural revolution which, around 10,000 years ago, provided the surplus food sources which made possible the great urban cultures of the ancient near east including those of Egypt, Ur and Palestine.[10] Archaeological findings reveal that many large settlements in this region were suddenly abandoned between the eighth and sixth centuries BCE, most likely because of the salinization of their water sources. Jeremiah lived after Isaiah and his extraordinary prophecies describe the gradual abandonment of some of the ancient cities of Israel. The prophet talks about the disappearance of water from the aquifers of Israel, and the effects of local climate change consequent on deforestation which showed in the disappearance of the permanent snows of the once deeply forested heights of Lebanon, in terms that are strangely reminiscent of the modern effects of climate change:[11]

> Does the snow of Lebanon vanish from the lofty crag?
> Do the proud waters run dry, so coolly flowing?
> And yet my people have forgotten me;
> they burn their incense to a Nothing.[12]

This moral and spiritual account of pollution does not end with the Old Testament. Jesus also mobilizes the language of pollution in his moral teachings although he is less concerned with physical pollution such as the ground water turning salty, and more concerned with an individual's spiritual condition. Like the prophets, Jesus argues that the true sources of pollution among the people of God were not their lack of devotion to dietary laws and sacrificial practices but their disordered desires, and their consequent failure to exemplify the justice and mercy of God in their relations with one another. Thus when he is challenged on his dietary laxity and that of his disciples he suggests that "it is not what goes into a person that defiles a person but what goes out of them." Jesus uses the analogy of roadways to indicate the peril which awaits those who pursue their desires unchecked; "broad is the gate that leads to destruction," while the "narrow way" characterizes life in the Kingdom (Matthew 7:13).

Jesus uses the language of moral purity and vision to contrast with the disordered gaze and polluting lusts of the unrighteous, as in the saying from the Beatitudes "blessed are the pure in heart for they shall see God" (Matthew 5:8). St. Paul similarly adopts the discourse of defilement and pollution, purity and righteousness in his letters to the Corinthians who, he argues, were endangering the unity of the body of Christ because they were polluting their bodies in sexual liaisons with prostitutes and in incestuous liaisons among the faithful. For Jesus and St Paul, as for Jeremiah and Isaiah, pollution named dangers to bodily health, ecological fertility and peaceable society among God's people whose origin lay in their moral and spiritual failings. We find this same recognition in Thomas Cranmer's language of "devices and desires" and "perils and dangers" in his *Book of Common Prayer* (1589). In the Christian moral tradition the human heart, not something called "nature," is the source of the greatest threat to human being and other than human flourishing.

## 2. The Discourse of Risk and the Discourse of Perils and Dangers

Modern accounts of risks from pollution are supposedly free of such moral or spiritual chains of causation. Pollution causes harms, according to the scientific narrative, because of physical causation. Carbon pollution threatens harms to human communities because it changes the climate, warms the oceans, causes floods and famines; biogeochemistry, not morality, is the authoritative frame. Human bodies are more prone to cancers in industrial societies because of the wide dispersal of novel chemically synthesized substances that industrial societies put in food, water and air; reptilians such as frogs and toads are especially sensitive to such pollution not because they are morally reprehensible but because their absorbent skins provide a more permeable barrier than those of mammals. But this cleansing of the modern narration of the human body and the earth from moral or spiritual logics does not mean that nature and pollution are somehow depoliticized. On the contrary, as Mary Douglas argues, whereas the ancients "politicized nature by inventing mysterious connections between moral transgressions and natural disasters" moderns politicize nature by selecting the dangers against which they will insure themselves, and by imposing environmental risks selectively on different social classes, groups or nations.[13]

Moderns narrate climate change in terms of the language of risk, and risk analysis involves the weighing of present benefits – from energy production and use – with present and future consequences of climate change. Risk analysis names a set of discourses and procedures which were borne out of the creation of the nuclear power industry.[14] Like anthropogenic climate change from coal, oil and gas burning engines and power stations, this form of energy production also involves threats to present and future generations. The biggest threat arises from the creation of long-term pollution in the form of indestructible and extremely life-threatening high level radioactive waste. Health threats from radiation are also intrinsic to the generation of nuclear energy. Cancer clusters among industry workers and their children, and in the geographical areas in which power stations are located, are extensively documented; and millions have been and will be affected by cancers from nuclear accidents such as those at Three Mile Island and Chernobyl.

The discourse of risk is quite different from the discourse of perils and dangers, not only because it lacks the moral and spiritual chains of causation which Jeremiah, Christ and Cranmer recognized, but also because the concept of risk carries quite different connotations from the word danger, originating as it does in the games of gamblers. The word risk, unlike the words peril and danger, indicates the probabilistic calculus of life in a technological society where the possibility of engaging in risky activities such as "unprotected sex" or driving an automobile without a seat belt, or at speed, is seen as evidence of human freedom from puritan moral codes or unduly rigorous laws. Modern risk analysis involves a set of practices where there exists no boundary between safety and danger.[15] The connection between gambling and the language of risk illumines because the most telling test that individuals apply to their personal arbitration of risks is related to choice. Most people are not disconcerted when they encounter the sometime negative effects of risks which they have chosen – injury from a skiing trip or an automobile accident, obesity from eating too much industrial food – but they are less happy when they encounter the negative effects of risks that have been imposed upon them, invisibly, by a public utility or a private corporation, such as heavy metals in drinking water or lead and nitrous oxide in the air.[16]

For many of those who are subjected to environmental risks choice often does not come into it as it is the poorest and least mobile who bear the greatest burden of such risks.[17] In modern America and Europe low income groups typically drink the most polluted water, breathe the foulest air and are more exposed to toxic substances in the workplace and in the home. This is because industrial plants, power generating stations, highway intersections, landfills and waste dumps are typically located in low income areas instead of adjacent to the homes and offices of the middle or upper classes, and because people at the bottom of the social hierarchy tend to do the dirtier jobs and inevitably encounter more toxic substances in their working lives.[18] Similarly on a biogeochemical scale it is the poorest of the earth's inhabitants, those who live at the tropics or in the Southern hemisphere, who are most at risk from the global contamination of the climate occasioned by the profligate use of energy by the richer countries of the North. Again the politics of pollution mirrors the dominant political structures and ideologies of the social world: the former colonial powers, their wealthy corporations, and the cities where they are headquartered, are less at risk from the threats of climate change than are the subsistence farms, coastal fishing villages and squatter settlements of those systematically impoverished by colonialism and postcolonialism in the Southern hemisphere.

In the matter of risk aversion, as in risk distribution, position in social hierarchies is also a major factor. Risk aversion is, unsurprisingly, linked with this perception of the inherently partial features of the social distribution of costs and benefits. Those who experience their societies as unjust are the most likely to suspect that the burdens or costs of nuclear power will not be shared in equitable ways.[19] And just like anthropogenic climate change, nuclear power involves present and future risks which are visited unequally on different groups of people.[20] The utilitarian and cost benefit rationale of those who use public taxes to finance nuclear power is that it is acceptable for society in general to enjoy the benefits of nuclear energy at the risk of particular individuals who are unable to escape its costs, such as people who work in or live near nuclear power plants, live or fish near waterways polluted by them, and consequently have a higher risk of cancer, or future people who have to bear the costs of present nuclear generated energy because of the indefinitely long periods over which its waste products must be stored and secured.

### 3. Involuntary Risk in Modernity and the Emergence of "Biopolitics"

Anthony Giddens suggests that risk in modernity is largely involuntary because globalization represents a social form where many interactions in daily life involve reliance upon persons and processes, corporate entities and governmental bureaucracies which are unknown and even invisible.[21] Modern human beings are increasingly dependent upon technical processes and corporate providers for the deliverance of food and clothing, heat and light, money and information, entertainment and education, and consequently life in modernity is one which involves faith and trust in persons and processes unknown. Constant reliance on the unknown involves a degree of trust in anonymous providers which produces a collective psychology of anxiety and distrust of socially imposed risks that is unique to modernity. This cultural condition produces a reaction which is manifest in "identity politics" and the quest for intimacy and individuation as the vital loci of faith and security in a world where so many procedures require a kind of procedural faith in large systems that are increasingly invasive and totalitarian. As Zygmunt Bauman argues, the nation state and the corporation use the threat of risk to justify their increasingly totalitarian powers over individual citizens and their local communities. After Karl Marx's famous aphorism "all that is solid melts into air." Bauman suggests that modernity specializes in making "*zuhanden* things into *vorhanden* things" by exposing "the fragility and unsteadiness of things" and "throwing open the possibility of (and the need for) reshaping them."[22]

Increasing quantities of greenhouse gases in the upper atmosphere are just the most extreme example of this larger tendency of the modern state to embrace instability and to impose new risks upon its citizens. The fears that these risks inculcate lend the modern nation state greater authority over its citizens who seem to need the state, through its administrative, legal and regulatory procedures, to defend them from the risks it imposes upon them. They also result in the emergence in the late modern era of "biopolitics"[23] for what was once perceived as the enduring stability of human nature arising from its divinely created character has been thrown, along with the rest of divine creation, into the melting pot. No more was it seen – no more could it be seen – as "given." Instead it has turned into a *task*, and a task which every man and woman had no choice but to face up to and perform to the best of their ability. "Predestination" was replaced with "life project," fate with vocation – and a "human nature" into which one was born with "identity" which one needs to saw up and make fit.[24]

Consequently the self becomes a distinctively ahistorical, asocial and individual project in late modern societies and the politics of the self are increasingly identified with decisions about diet, exercise, clothing, sexual identity, and even technological manipulation of the body through plastic surgery or genetics. As Bauman puts it "no more does one's place in society come as a, wanted or unwanted, gift" for "needing to *become* what one is, is the feature of modern living."[25] In constructing personal identities through such multiple decisions individuals are conscious that they are also interacting with a growing level of uncertainty and risk. Modern individuals are no longer pilgrims who know where they are going and are seeking how to get there, but "vagabonds" whose task "is to pick the least risky turn at the nearest crossroads, to change direction before the road ahead gets impassable, before the road-scheme has been redesigned, or before the coveted destination has been moved elsewhere or lost its glitter." This new politics of risk in a social order which is constantly being torn up and remade by the procedures and institutions of late capitalism interacts with genuine environmental threats, such as the threat of climate change. But such threats are linked to the decisions and policies of corporations and governments, public services and utility companies, even terrorists, over which individuals have appearingly no control and yet whose decisions to continue producing greenhouse gases, or to put newly synthesized chemicals or known toxins into food or water, air or ocean, affect the health and safety of billions of people. Life politics consequently dominates the political processes of late modern democracies while collective fear and anxiety become the characteristic cultural condition of late modernity.[26]

## 4. Fear or Reverence as the Foundation of Christian Ethics

In the Christian moral tradition fear is understood as arising from a moral or spiritual lack instead of as the basis for moral deliberation. But some argue that fear of the uncertain consequences of climate change should be the prime motivator for collective action on the issue. John Houghton, former chair of the IPCC, suggests that fear of climate ought to be stronger than fear of international terrorism for it represents a more certain threat to human security. But while it may be true that fear of the cataclysmic effects of climate change helps to explain the possibility of international agreement on the Kyoto Protocol, to which 187 nations are signatories, nonetheless fear of our demise cannot form the basis of a Christian ecological ethic if the form of the moral life of the Christian is to mirror the biography of the God-Man Jesus Christ who did not count his life worthy to be saved but "who gave his life a ransom for many" (Matthew 20:28). As Stanley Hauerwas argues fear of death or a fearful struggle for survival do not belong in a Christian moral frame.[27] Reverence, not fear, is the foundation of Christian ethics, reverence for God and reverence for God's creatures. And the profligate use of the planet's million year store of carbon in the last three hundred years manifests above all else a profound lack of reverence toward God and God's earth.

In his prophetic essay *The Abolition of Man*, Lewis argued that the combined effects of astronomy and chemical agriculture had evacuated divinity from the human view of the earth and the universe and that this was why modern human beings look within, to their inner lives and emotions, for the truth of their existence. They are trained by modern philosophers no longer to see what they nature as morally or spiritually meaningful, and as the physical world is rendered meaningless, so modern human beings are delivered up to inner experience and emotion as the arbiters of moral order. But the irony is that this new moral sentimentality renders moderns peculiarly vulnerable to totalitarian and technological threats to their flourishing because it renders the physical and social worlds peculiarly amenable to the hubris of human control and dominion. As Dietrich Bonhoeffer argued in his reflections on Genesis 1-3, *Creation and Fall,* it is because we "cease to know the world as

God's creation," that "the earth is no longer our earth, and then we become strangers on earth," and from strangers we finally become earth's subjects: through the power of technology "the earth grips man and subdues him."[28]

The threat of catastrophic climate change is the ultimate instance of this grip of the earth we have failed to revere, and this failure of reverence is, as Lewis and Bonhoeffer also saw, deeply connected to the modern failure, since the Enlightenment, any longer to regard the earth as a good earth, or as God given. Far from this being the best of all possible worlds moderns are trained to see the world as chaotic and violent, and full of threats to human life from which only the leviathan of the modern state and the alchemy of modern science can deliver it. The good life for humanity on this view is only deliverable in the extent to which humanity controls the chaotic and meaningless elements of the natural and social worlds. Reverence for God and for divine providence is supplanted in modernity with the sovereignty of reason and the power of science to deliver human beings from fear of scarcity and war. But as we have seen the modern nation state generates collective fear of the risks it imposes in order to legitimate its collectivist endeavors to engineer human lives and even the planetary systems. And the threat of climate change, as it is narrated by politicians and scientists, appears to pose no challenge to this world view. On the contrary many scientists, backed by corporations and public bodies, are actively pursuing technological solutions to climate change: such solutions include efforts to find ways to reflect back more of the sun's rays from the surface of the earth to compensate for global warming, and efforts to enhance the planet's capacities to absorb carbon dioxide pollution, such as genetically modified forests, or oceans modified with iron filings so they grow more plankton and algae. Control is the primary motif of the characteristically modern economic, political and science-informed institutions of the nation state and the corporation, just as it is the prime character trait imbued by the daily rituals of energy consumption. Risk analysis is the discourse and set of procedures which governments have invented in order to seek to control the dangerous consequences, as well as to legitimate the heavy burden, of such novel technologies as nuclear power and genetic engineering.[29] But the frame of risk analysis does not challenge the underlying motif of control over the forces of nature, and over the citizenry, that the modern nation state and its corporate partners claim in the name of the project of limitless energy use and infinite consumption; on the contrary it reaffirms it in a new guise.

## 5. Future Generation, the Inadequacy of Consequentialist Calculations, and Christian Eschatological Hope

Compared to the known risks, and long run costs of nuclear power, the costs and risks of anthropogenic climate change remain relatively unknown. In a situation of uncertainty prudence would seem to be even more advisable than it is in the case of known consequences. Aristotle called consideration about consequences and future uncertainty "deliberative discourse."[30] Deliberative discourse involves uncertainty because it involves judgments about future circumstances which are unknown in the present. Therefore in judgments about the future, more than in judgments about the past, the credibility of the speaker is crucial, because if a person is perceived as fair-minded then he or she is more likely to be trustworthy in deliberations on future situations which include several unknowns.[31] Hence in situations of future uncertainty *ethos* is central to the moral authority of the one who claims to be able to deliberate: if a person is known to be practiced in the art of *phronesis,* or practical reasoning, and lives the life of the virtues, then their good character and practiced wisdom will make their predictions more persuasive.[32] Aristotle's argument shows why it is that consequentialism on its own is such an inadequate basis for the moral life of a community or a polity.

Peter Singer is the most consistent of contemporary advocates of consequentialism in that he applies the consequentialist logic equally to human beings and other animals. Singer suggests that it is just because moderns no longer find any correlation between the way they ought to live and the structure of the universe that estimations of the consequences of their actions for the welfare of other beings are all that they have for moral guidance.[33] This consequentialist calculus involves him arguing for killing "defective" babies, and it is the same calculus which enables millions of individuals every year to terminate the lives of their unborn children. Climate change also threatens the unborn. But how likely is it that a civilization which kills so many potential people every year will be able to give moral suasion to the effects of its energy policies on presently unborn people? It is the case that anthropogenic climate change has led to a rash of scholarly papers on the moral problem of future generations, though there exists little evidence so far that estimates of the possible consequences of present actions for future people who do not currently exist are morally persuasive to consumers, corporations and governments.[34]

The reason that future people count for so little in modern moral deliberation is because of the modern assumption that the future is empty, up for grabs, and that collectively human beings make their future and even the future of the planet.[35] In Christian eschatology the future is perceived as already given in the events of the coming of Christ and the foretaste of the Kingdom which already exists in the new creation which God in Christ inaugurates. However, since the Enlightenment moderns have been trained to believe that the future is not given by God but the outcome of the choices and projects of humanity, an approach which is sustained by modern devotion to the idea of progress. Paradoxically this is an idea which orients modernity toward the future while at the same time it enables consequentialists to discount costs and risks for future peoples against present benefits for existing ones, on the assumption that future peoples will be able to sort out the risks run by present ones because of "progress."

There exists an underlying problem with time here which helps explain the puzzling inadequacy of consequentialist calculations in relation to posterity: consequentialists are not so much concerned with the *now* of the ethical moment or the present goodness or badness of a particular act as with the state of affairs which will pertain *after* the event or act in question. The consequentialist looks to the ends of actions, instead of the intrinsic character of the actor or the action. As Philip Pettit puts it under consequentialism "agents are required to produce whatever actions have the property of promoting a designated value, even actions that fail intuitively to honour it."[36] This is why consequentialists reject judgments about actions on the basis of how they have been construed in past moral traditions. The origin of this attempt to eschew time past from moral rationality may well be Immanuel Kant's construction of the moral agent as an innocent with no past or future in his *Religion Within the Limits of Reason Alone*, an approach which finds more recent analogy in John Rawl's idea of the "original position" by which Rawls intended to clarify a concept of justice as social contract untainted by personal interest, memory, narrative or tradition.[37] The attempt to construct a version of moral rationality which is timeless except inasmuch as it looks to the future makes self-alienation a central feature of consequentialist ethics because moral reasons are only possible to individuals who view their moral projects as outside observers, as if they were not valuable to them.[38]

Against the forgetful and unsituated consequentialist calculus of modern ethics, Christian ethics is intrinsically connected with memory and tradition, for Christians believe that their future and the future of creation is already given and revealed as the "new creation" which God inaugurates in the events of the life, death, resurrection and ascension of Jesus Christ as narrated in the Christian scriptures. For Christ is "before all things" (Colossians 1:17), "he is the beginning" (Colossians 1:18), "through him God was pleased to reconcile to himself all things, whether on earth or in heaven, by making peace through the blood of the cross" (Colossians 1:20) and in him God promises to fulfill the creation at the end of time when Christ will come again and "gather up all things in him" (Ephesians 1:10). For Christians the memory of these events and of this God-given future forms the basis of moral deliberation. Christians do not presume to know in advance the outcome of their actions except inasmuch as they remember that God has already determined their end, and the end of all things in the Resurrection of Jesus Christ from the dead.[39] And this is the spring of hope and moral courage for Christians, who live in expectation that their lives, being caught up by grace in the life of the Spirit, may manifest the goodness of God even in the midst of their brokenness and sinfulness.

## 6. The Real Energy Crisis and Christian Truthful Speech

In the perspective of Christian eschatology the profligate use of energy in industrial societies is a moral peril not primarily because of present or future threats to human survival that anthropogenic climate change may cause but because of the kinds of people, the kind of planet and the kind of society fostered by the goal of infinite consumption, and the extraction and consumption of energy without limits to which this goal directs modern civilization. As Wendell Berry puts it, the real energy crisis is not that occasioned by climate change or the unavailability of oil in the near future but the threats to the good life and the good society occasioned by the habitual dependence of modern consumers on the systematic waste of the earth's energy reserves in the modern economy. The energy hungry systems of modern agriculture and industrialism subvert the health of the soil and the community between soil and soul on which good agriculture and good living depend. Energy and machine dependence subvert the kinds of characterful lives and practices that make the virtues of prudence and temperance, justice, courage and hopefulness habitual and which enabled farmers and producers in previous eras to grow food, heat their homes and journey from place to place without destroying the earth or threatening the welfare of future generations.[40] In this perspective the risks that human beings face in relation to their use of energy are as much moral and spiritual as they are climatological or ecological; the free availability of energy to fuel our comfort-zoned lifestyles with their daily dependence on machines and entertainment devices produces people and lives which lack depth and vigor. It also fosters the dangerous illusion that our choices, and not the gifts of God's good earth, determine our destinies as individuals and as a species.

The conventional narrative of climate change indicates that it represents a potential threat to persons – subsistence farmers in Bangladesh or future generations – who are mostly so distant from the profligate users of energy that the consequences of excess hardly seem a moral priority. But if Berry is right the moral judgment that should move us most concerns not these distant and so barely perceived consequences but that our energy dependent lives will harm not just future people but soil and soul in our lands and present lives. Biogeochemists and climatologists imagine that the threat of future physical harms from climate change, provided they can be described with the minimum of uncertainty or vacillation, should be enough to persuade the population of the need to abandon its energy-hungry ways. But the scientific description of climate change is not persuasive because it does not describe the real moral and spiritual perils represented by limitless energy consumption, or the modern project of the infinite consumer and the modern extractive economy which turns nature into a mine and a waste dump at one and the same time.[41] The infinite consumer is the product of the disordered desires manipulated by the psychological informed marketing and advertising industry.[42] But she is also the creation of an industrial economy which from its inception rested upon the release from the crust of the earth sources of energy which knew no bounds; limitless growth becomes the addiction of modernity and this is why, ultimately, climate change deniers continue to deny that climate change is a moral imperative. So overpowering is the modern economic account of the good society as one of endless growth which requires the availability of unlimited supplies of energy to feed the unlimited desires of the infinite consumer that it becomes impossible to imagine an alternative vision of human flourishing, and especially one which acknowledges the limits of the planet to absorb the industrial waste of its energy store.

In some of his last lectures, given at University of California, Berkeley, in 1981, Michel Foucault suggested that those who claim to be the arbitrators and authorities in the matter of risks and future uncertainties typically use their special knowledge as claims to power and control. But such power claims while they may be persuasive are not truthful for they mask the identity of the speaker and the true character of the context from which they utter. Foucault proposes that the quest for truth back of such claims to power involves the ancient ethical practice of *parrhesia* or "fearless speech," which is another way of speaking of what Christians and Jews call prophecy.[43] In the art of *parrhesia* the truthfulness of the utterances of fearless speakers resides in part in their willingness to speak less on behalf of others than alongside them. Truthful speech which takes the form of the ritual of confession is speech which is powerful enough to produce modifications in the hearer *and* in the speaker because it "exonerates, redeems, and purifies him; it unburdens him of his wrongs, liberates him, and promises him salvation."[44] Truthful speech is not only speech which acknowledges wrongs done but which acknowledges a share in the burden of wrong and a preparedness to relent, and so makes possible forgiveness and reconciliation. On this account fearless speech requires a ritual context in which it can be heard and acknowledged; it also requires the courageous compassion of prophetic individuals like Hosea or Socrates, Christ or John the Baptist, who are prepared not only to acknowledge but to bear the burden of their truthful utterance. The truthful speech is truthful because of the preparedness of the speaker to lay down her life for her friends.

In a well-known essay Stanley Hauerwas renarrates the role of truthful speech in Richard Adam's remarkable novel *Watership Down.*[45] Hazel is the fearless rabbit whose friend Fiver perceives in a real estate notice pinned to a tree a coming cataclysm which will destroy the warren of Sandleford. Hazel approaches Silverwood, the leader of the warren, but is dismissed as a doom-monger and so Hazel and a small band of followers embark on a long and arduous journey at the end of which they eventually found a new warren. Along the way they make mistakes but through it all they remain true to their original heroic insight which was heroic not because they were strong but because they had the courage to act on it despite their not being strong. Like Frodo in the Lord of the Rings, Hazel is an unlikely hero for he has no intrinsic strength except his preparedness to rely upon his friends, and even on strangers, in bearing the burden of truth, and his preparedness to give up the known for the unknown, a comfortable life in order that others might find life again.

Climate change threatens to disturb our planetary home, just as much as the real estate development threatened the warren of Sandleford. But Christians cannot expect that the consequentialist procedures of power hungry institutions will produce the kinds of truthful utterance which is heard as shared confession precisely because for them so to do these institutions would have to acknowledge the lies, the systemic injustices, and the irreversible ecological destruction on which they are built. The truth is that as things look now it is not the homes of those who treat the earth's energy as a limitless resource which are first threatened by climate change but the homes of others who are far less able to defend themselves. *How* the story of climate change is narrated therefore crucially determines the kinds of moral responses that are made to it. In the end this ultimate risk to planet earth poses the question of what kind of people we aspire to be, and what is the vision of the good society to which our lives and practices commit us? The procedures and institutions of international environmental treaty are already grinding out an answer to the problem of climate change but in the form of the Kyoto Protocol it is an answer framed entirely in the probabilistic calculus of consequences. And it is an answer which favors the powerful over the powerless, and which continues to find ways of justifying present profligacy over future risk, and which fails to challenge the dangerous illusion which our energy dependent lives sustain that "we" are in control.

## NOTES

1. Thomas Cranmer, *Book of Common Prayer* (London: 1589).

2. World Health Organisation, *Climate Change and Human Health: Risks and Responses* (Paris: World Health Organisation, 2003).

3. Intergovernmental Panel on Climate Change, *Synthesis Report* (Cambridge: Cambridge University Press, 2001).

4. IPCC, *Synthesis Report*. See also Jeremy Leggett, *The Carbon War: Global Warming and the End of the Oil Era* (New York: Routledge, 2001), pp. 14□19.

5. William H. Schlesinger, "Better Living through Biogeochemistry," *Ecology*, 85 (2004), pp. 2402□2407; also Eville Gorham, "Biogechemistry: Its Origins and Development," *Biogeochemistry*, 13 (1991), pp. 199□239. See further James Lovelock, *Gaia: A New Look at Life on Earth* (Oxford: Oxford University Press, 1979).

6. Richard B. Alley, *The Two-Mile Time Machine: Ice Cores, Abrupt Climate Change, and Our Future* (Princeton NJ: Princeton University Press, 2000).

7. Genesis 9:12.

8. Robert Murray, *The Cosmic Covenant: Biblical Themes of Justice, Peace and the Integrity of Creation* (London: Sheed and Ward, 1992).

9. Margaret Barker, "The Book of Enoch and Cosmic Sin," *The Ecologist* (January, 2000).

10. On the origins of the city in the ancient near east see further Lewis Mumford, *The Culture of Cities* (London: Secker and Warburg, 1938).

11. Jack V. Thirgood, *Man and the Mediterranean Forest: A History of Resource Depletion* (London and New York: Academic Press, 1981).

12. Jeremiah 18:14□15.

13. Mary Douglas and Aaron Wildavsky, *Risk and Culture: An Essay on the Selection of Technical and Environmental Dangers* (Berkeley: University of California Press, 1982), p. 30.

14. Carolyn R. Miller, "The Presumptions of Expertise: The Role of *Ethos* in Risk Analysis," *Configurations*, 11 (2003), p. 164.

15. Zygmunt Bauman, *Postmodern Ethics* (Oxford: Blackwell, 1993), pp. 199□200, note 14.

16. Douglas and Wildavsky, *Risk and Culture*, pp. 16□21.

17. Mary Douglas, *Risk Acceptability According to the Social Sciences* (London: Routledge and Kegan Paul, 1986), p. 6.

18. Robert D. Bullard and Beverly H. Wright, "Environmental Justice for All: Community Perspectives on Health and Research Needs," *Toxicology and Industrial Health*, 9 (1993), pp. 821□841.

19. Stanley Rothman and Robert S. Lichter, "The Nuclear Energy Debate: Scientists, the Media and the Public," *Public Opinion* (1982), pp. 47□52 cited Mary Douglas, *Risk Acceptability According to the Social Sciences* (London: Routledge and Kegan Paul, 1986), p. 6.

20. Constantine Hadjilambrinos, "An Egalitarian Response to Utilitarian Analysis of Long-lived Pollution: The Case of High-level Radioactive Waste," *Environmental Ethics*, 22 (2000), pp. 43□62.

21. Anthony Giddens, *The Consequences of Modernity* (Cambridge: Polity Press, 1990) p. 25.

22. Zygmunt Bauman, "Identity in the Globalising World," *Social Anthropology*, 9 (2001), pp. 121□129, p. 122.

23. See further Nikolas Rose, "The Politics of Life Itself," *Theory, Culture and Society*, 18 (2001), pp. 1–30.

24. Bauman, "Identity in the Globalising World," p. 122.

25. *Ibid.*, p. 124.

26. Judith Shklar, "Liberalism of Fear," *Liberalism and the Moral Life*, ed. Nancy Rosenblum (Cambridge, Mass.: Harvard University Press, 1989).

27. Stanley Hauerwas, "The Moral Limits of Population Control," *Truthfulness and Tragedy: Further Investigations into Christian Ethics*, eds. Stanley Hauerwas with Richard Bondi and David B. Burrell (Notre Dame, Ind.: Notre Dame University Press, 1977).

28. Dietrich Bonhoeffer, *Creation and Fall: A Theological Interpretation of Genesis 1–3*, trans. John C. Fletcher (London: SCM Press, 1959), p. 38, and also Michael S Northcott, *The Environment and Christian Ethics* (Cambridge: Cambridge University Press, 1999).

29. Harold W. Lewis, *Technological Risk* (New York: Norton, 1990) cited Miller, "The Presumptions of Expertise," p. 166.

30. Aristotle, *On Rhetoric: A Theory of Civic Discourse* I. iii. 5, cited Miller, "Presumptions of Expertise," p. 167.

31. Miller, "Presumptions of Expertise," p. 167.

32. *Ibid.*, p. 167.

33. Peter Singer, *How Are We To Live? Ethics in an Age of Self-Interest* (London: Mandarin, 1994), p. 188.

34. Hauerwas, "The Moral Limits of Population Control."

35. William Schweiker, Responsibility and Christian Ethics (Cambridge: Cambridge University Press, 1995), p. 129.

36. Philip Pettit, "Consequentialism," *A Companion to Ethics*, ed. Peter Singer (Oxford: Blackwell, 1993), p. 231.

37. John Rawls, *A Theory of Justice* (Oxford: Oxford University Press, 1973), pp. 118–142.

38. Stanley Hauerwas and David Burrell, "From System to Story: An Alternative Pattern for Rationality in Ethics," *Why Narrative? Readings in Narrative Theology*, eds. Stanley Hauerwas and L. Gregory Jones (Grand Rapids, MI: William B. Eerdmans, 1989), pp. 158–190.

39. Douglas Farrow, *Ascension and Ecclesia* (Edinburgh: T and T Clark, 2000), p. 214.

40. Wendell Berry, *The Unsettling of America: Culture and Agriculture* (San Francisco, CA: Sierra Club Books, 1977), pp. 81–84.

41. On the concepts of the infinite consumer and the extractive economy see further Crawford B. Macpherson, "Democratic Theory: Ontology and Technology," *Democratic Theory: Essays in Retrieval*, ed. Crawford Macpherson (Oxford: Oxford University Press, 1973), pp. 24–38.

42. Macpherson, "Democratic Theory." See also Timothy Gorringe, *The Education of Desire* (London: SCM, 2001).

43. See further Michel Foucault, *Fearless Speech* (Paris: Semiotext, 2001) cited in Nancy Luxon, "Truthfulness, risk, and trust in the late lectures of Michel Foucault," *Inquiry*, 47 (2004), pp. 464–489.

44. Michel Foucault, *History of Sexuality,* Vol. 1, trans. Robert Hurley (New York: Vintage Books), pp. 61–62 cited Luxon, "Truthfulness, Risk and Trust," p. 475.

45. Stanley Hauerwas, “A Story-Formed Community: Reflections on *Watership Down*,” *A Community of Character: Toward a Constructive Social Ethic* (Notre Dame, Ind.: University of Notre Dame Press, 1981), pp. 9–35.

# Six

# GLOBAL ETHICS, ENVIRONMENTALLY APPLIED: AN ISLAMIC VIEW

Anis Ahmad

Realization of ethical and moral behavior, according to the Islamic scripture, is the core issue for which the ultimate Creator of humankind sent messengers, one after another, in different parts of the world and in different periods of history. They shared in one common concern, namely, that human beings should act virtuously and fight against evil ways of doing things. "We raised a Messenger in every community [to tell them] to serve Allah and shun the evil one."[1] The role of Prophet Muhammad and earlier Messengers of Allah is defined in strict ethical terms as realization of *ma'ruf* (good and virtue) and eradication of *munkar* (wrong, unethical behavior). Luqman advises his son, "Son, establish Prayer, enjoin all that is good (*ma'ruf*) and forbid all that is evil (*munkar*), and endure with patience whatever affliction befalls you."[2]

This concern for ethical conduct is perhaps also the most profound common thread in the living world religions. While there may be apparent differences at a micro level, in general, all world religions teach honesty, truthfulness, fidelity, peace, love, and respect for fellow human beings and fairness to their followers.

Appropriate knowledge and internalization of ethical values help in peaceful coexistence between people of the world, while lack of proper inculcation of ethical values leads to religious violence, extremism, intolerance, and conflict.

Islam, youngest in the Abrahamic tradition of faith, advocates a wider role of ethics at personal, social, and environmental levels. The ethical and moral teachings of the Qur'an and the Prophet offer general principles, specific instructions, commands, and directives in personal matters, social policy, and global ecological issues. A direct and closer look into the contents of the Qur'an and the Prophetic *hadith* and *sunnah* (statements, actions, and endorsements) reveals that theological issues do not occupy a central place in the thematic scheme of the Qur'an. There is hardly a page of the Book where ethical and moral conduct in social, economic, political, and legal matters is not underscored.

Environmental balance and purity are linked directly, in the Islamic world view, with the people's way of living. Honest and fair behavior of individuals leads to a healthy environment, ecological balance, and moderation in their lifestyle. Misuse of resources and the practice of injustice and inequity invite ecological and environmental crises and catastrophes.

This chapter tries to identify the scriptural foundations, from an Islamic perspective, for an ecologically friendly world. It also explores the need for a global ethical force that may create awareness among people of diverse faith and culture on ecological issues. We also try to understand the nature of the relationship between ethical norms and values of a people and their ways of treating fellow human beings and environmental harmony and balance.

## 1. Globalization, Economic Imbalance, and the Environmental Crisis

One major benefit of globalization is the fast and easy access and sharing of information, thanks to global networking of information. Geographical boundaries have virtually ceased to exist because of the free travel of information all over the globe. Electronic commerce, e-government, and e-money have reduced the traditional world to a global village. Globalization has its direct bearings on education, media, society, and economy. One major economic challenge of globalization is that in order to survive and meet their needs, most of the developing countries will have to double their exports to the developed world. The developed world will, in turn, get these products at a comparatively low price, even when the exports from the developing world are doubled. However, this will not make much change in their poverty level. While doubling their exports, the developing nations are compelled to be unfair to their environment, as we see in the case of South American countries.

Brazil, a country with thirty percent of the planet's rain forests, long considered to be "the lungs of the world," is facing an environmental crisis thanks to the globalization of the economy. In 1999, due to pressures of the International Monetary Fund (IMF) the country reduced its budget on environmental spending by nineteen percent; the consequences were obvious. In Indonesia, according to an estimate, the country's forests are hacked down at the rate of six million acres (2.4 million hectares) per year. In the Malaysian state of Sarawak, thirty percent of the forests disappeared in only two decades, while seventy-three percent of 116 rivers surveyed by the authorities were found to be either "biologically dead" or dying.[3]

In Madagascar, according to an estimate, the forest is vanishing at the rate of 500,000 acres (200,000 hectares) a year. Poverty is obviously one major source of the problem as seventy percent of the island's fourteen million people live on less than one dollar a day. This unjust distribution of wealth and ever-increasing control of the rich nations over the economic resources of the world, in the name of globalization, compelled 50,000 people concerned with the environment from dozens of countries of the world to gather in Seattle in November 1999 and protest at the annual meeting of the World Trade Organization (WTO). In April 2000, another 15,000 protesters and environmentalists gathered in Washington, D.C., to protest against unfair policies of the IMF and the World Bank for the same reason. Concurrent to it is the rise of unilateralism, which, in the name of a new capitalist world order, imposes on the rest of the world economic and political hegemony of one great power. The so-called New World Order (NWO) also tries to impose in the name of globalization one brand of political order, fast food outlets, and Hollywood culture.

In addition to the socio-political and cultural threats, namely, the imposition of WTO policies and the doctrine of "regime change," we are also facing a growing danger of a global environmental crisis. At the root of this crisis are two apparent factors. First is the emergence of a capitalist economy which creates mountains of waste and eats up green forests. According to available data on the Internet, it takes 75,000 trees to print the Sunday edition of the New York Times alone. One can imagine how many green forests are annihilated, not only every week, but throughout the year by only one symbol of modernity. The plight does not end with the printing of the New York Times, Los Angeles Times, and Chicago Tribune, to name only three major U.S. newspapers.

Annually, Americans throw away office and writing paper that would be enough to build a twelve-foot-high wall stretching from Los Angeles to New York City. According to U.S. sources, Americans buy sixty-two million newspapers and throw away around forty-four million every day. That is the equivalent of dumping 500,000 trees a day. Similarly, in his or her lifetime, the average American throws away 600 tons of garbage. This means that each adult will leave a legacy of 90,000 pounds of trash for the next generation. Moreover, the quantity of waste Americans produce every three months by throwing away aluminum cans used for soft drinks is enough to rebuild an entire commercial aircraft. If users in the U.S. would recycle just one-tenth of the newsprint that is currently being thrown away, we could save an estimated equivalent of about twenty-five million trees a year. The United States is not alone in such massive generation of waste. It is perhaps an attribute of most of the industrialized world having high-profile consumerism. Population growth is another important factor that has direct relationship with the environmental crisis, thanks to increased demand of food, fuel, water, fiber, timber, and land. The socio-economic imbalance thus created has led to a widening gap between the poor and the rich at a global level.

Second, a recent report that involved experts and scientists from ninety-five nations, including United Nation agencies, warns that "a rising human population had polluted or over-exploited two-thirds of the ecological systems on which life depends, ranging from clean air to fresh water, in the past fifty years."[4]

This is causing enormous strain on the natural functions of Earth and on the ecosystems of the planet needed to sustain future generations. According to an estimate, twenty-four percent of the land on Earth is already cultivated. Since 1980, twenty percent of the world's coral reefs have been destroyed.

Water withdrawal from lakes and rivers has doubled in the last forty years. Today, humans use between forty and fifty percent of all available fresh water.

It is generally feared that the pollution and contamination in many parts of the world have reached alarming levels. Most of such warnings assume the increase in population in the developing world as a major contributing factor to the environmental crisis. However, it is interesting to note that the developed nations, which have been able to contain and control their population, and not the developing world, are mostly responsible for the pollution of air and water thanks to their massive use of chemicals, the emission of $CO_2$, and food demands. People in Asia and Africa, though outnumbering Europe and America in population, are more a victim of pollution caused by over exploitation of resources by the so-called developed world. Large-scale industrialization and luxurious lifestyle in the capitalist world have added to this crisis.

On a closer analysis, the view that the developing world should contain the size of its population appears perhaps as part of a strategy of the developed world to allow itself to enjoy an extravagant lifestyle at the cost of the developing and poor nations.

The industrialized world, more specifically Europe and North America with their shrinking populations, have an alarming rate of consumption of a variety of food items as compared to a very austere and bare minimum utilization of food by the South. The water resources of developing nations are also systematically brought under the control of the rich nations.

This does not mean that the European and American consumer society alone is to be blamed for many environmental irregularities committed by the Asian and African nations. This only indicates that the contention that comparatively high birth rates in Asia and Africa are the key factors responsible for the environmental crisis, is not fully supported by the reality on the ground.

Apparently, air pollution, contamination of water, increase in loss of land resources, and shortage of natural resources are the core issues in most of the contemporary environmental studies. An important factor, not properly highlighted in our view, is the extravagant use and unfair division of available resources. This is not due to physical rape of the environment but to the mind-set, the normative framework, and the ethical and moral vision of the corporate world and in the end to the users in a materialist-capitalist society. If the psyche and moral vision of the members of a society is self-centered, ego-centric, and individualistic, the environmental crisis becomes inevitable.

## 2. Islamic Ethical World View

The human being remains the key player and central actor in the contemporary environmental discourse. It reinforces the view that not only our material needs but also our ethical and moral vision have direct relevance in one's encountering the environment. A human's physical, mental, emotional, cultural, and social needs are directly related to ethical values. Since the capitalist society and economy has its philosophical foundations in individualism, utilitarianism, and materialism, its business ethics is also a reflection of these principles. Their application in society consequently causes imbalance and inequities.

Ethics and morality, in the context of the Islamic world view, remain the basic and primary categories. The human being, in the Islamic scheme of life, is not a mere social animal; "human being" is an ethically and morally responsible being. For instance, Islam requires that uprightness, fairness, and truthfulness be observed by the individual, not only in personal matters, but in his or her socio-economic conduct, irrespective of space and time.

Being applied in its nature, Islamic ethics and morality carry important ecological implications. It guides an individual, society, and state in how to relate itself economically with fellow humans, with the cosmos, and with the Creator and Sustainer of the cosmos.

It is interesting that, instead of dilating on theological issues, the Qur'an invites human reason and intellect to think critically on environmental and ecological issues and ultimately guides its student to develop *an ethically-motivated behavior.* "Indeed it is He who sends the winds as harbingers of auspicious news announcing His beneficence, bringing heavy clouds which we drive towards a region lying dead and send down rain, and raises all kinds of fruit. So shall we raise the dead, that you may think and reflect."[5]

Referring to the observable environmental phenomenon with which humans, whether in deserts or in cosmopolitan cities, are familiar, the Qur'an invites "human beings" to reflect and to use tools of reason in finding out how wind carries clouds with massive amounts of water to far-off places in order to give new life to barren lands. Frequent references to the environment, including water, air, and land, motivate the human mind to understand and analyze the ecological systems in the universe. It also leads the human imagination to draw lessons from the observable effects of rain on dead land, and to visualize how humankind will be resurrected one day to face a moral and ethical judgment.

## 3. Cosmic Balance

Reminding the human being of the love, compassion, and concern of the Creator for His creatures, the Qur'an says, "It is He who sends down water from skies and brings out of it every thing that grows, the green foliage, the grain lying close, the date palm trees with cluster of dates, and the gardens of grapes and of olives and pomegranates, so similar, yet so unlike. Look at the fruit, how they appear in the trees when they ripen. In all these are signs for those who believe [and reflect]."[6]

The environmental message conveyed in this passage, which is repeated elsewhere in the Qur'an, is simple. "Human being" should respond to these favors of Allah, the Creator of humankind and the universe by developing a responsible, reflective, and pro-active attitude with ecological consciousness. It also reminds the human being, as an actor in history, to benefit from the favors and bounties of Allah, the Creator, and not to be excessive. "And those who, when they spend, neither too lavishly nor yet hold back and there is ever a just balance between the two,"[7] "and be not extravagant Allah loves not the prodigals."[8] Balance, moderation, and an ecological consciousness thus emerge as a major behavioral objective of humankind.

The Qur'an also talks about mountains with their roots deep in Earth as a balancing device and as a pressure on the crust of Earth against the molten substances deep in its center. "We stretched the earth and placed upon it firm stabilizers (mountains) and made all things grow upon it balanced evenly."[9] In referring to the mountains as stabilizers, the Qur'an refers to the ecological balance that exists at a geological level as food for thought for human being. Those who have an objective and an unbiased mind are thus repeatedly invited to recognize the One who is behind this ecological balance – the One who sustains life and also promotes it.

The Qur'an invites a perceptive and critical mind to investigate and explore the environment in order to understand and appreciate the ecological relation which exists between the human being and universe.

Its references to the environment are remarkably frequent. On almost every page of the Qur'an, in one context or another, it uses terms that invite humans to observe and reflect. For example, the word "Earth" *(ard)* appears no less than 485 times in the Qur'an. Reference to water is made in around 300 places, and the word "sky" is used in near fifty places in the Qur'an. Similarly, the term *shari'ah* literally means "source of water" or "stream" and at the same time means laws provided through divine revelation *(wahi).*

While reminding humankind of God's favors, such as timely rain showers, fresh air, variety of fruit, and so on, the Qur'an invites humans to develop a conscious moral attitude of thankfulness toward their Benefactor. A thankful attitude and behavior are defined further by the Qur'an as an environmentally friendly conduct that makes a conscious effort to live a balanced, moderate, and simple life. Lavish and extravagant use of resources is regarded as an unthankful attitude. Consequently, the Qur'an discourages rather than condemns lavish and wasteful use of natural resources.

While dilating at a conceptual level, the Qur'an addresses itself to the root causes of the environmental crisis and imbalance. One major cause, as the Qur'an points out, is inequitable and unjust distribution of resources, which are given to humankind as a trust (*amanah*) by the Creator. Similarly, the ethical and responsible attitude in consumption is underscored by the Qur'an and the *sunnah*. In a remarkable piece of wisdom, the Prophet (peace be upon him) advises that even when a person makes ablution at the bank of a river, one should not wash face, hands, or other parts of the body more than three times.[10] This simply means that even when one has access to abundant resources, he or she should act with responsibility, moral constraint, and austerity.

The Prophet, in another of his sayings, persuades his followers to plant trees as an ongoing charity *(sadaqah jariyah).* He instructs his followers to plant in order to be rewarded with as much return as it yields. One is advised to plant a tree even when signs of the end of the world or the falling of the day of judgment have started appearing. Similarly, when sending out troops, the Prophet specifically instructs them to not destroy trees, burn orchids, or plunder farmlands. A *hadith* (saying) of the Prophet specifically prohibits the believers from polluting ponds, lakes, and waterways by throwing into them human refuse or other impurities and waste.

The Prophet instructs his followers to not only protect the environment and natural resources but also to conserve animal life. A *hadith* says, "If a person kills a sparrow unnecessarily, he will be questioned for it in the day of judgment."[11]

## 4. Ethically Motivated Behavior

Islamic environmental ethics goes beyond the conventional discourse of land, air, and water pollution generally caused by industrialization. It takes serious note of de-forestation, over-fishing, putting chemical waste in oceans thus causing the death of coral reefs, hunting of animals and birds to the extent of endangering their species, noise pollution, and so on. However, its primary focus remains on the root cause of these destructive actions. Relating physical manifestations of pollution and inequity with the inner imbalance of personality, attitude, and behavior pattern of humans, it highlights the role of the inner self. The Qur'an suggests that humans essentially have good nature thanks to their good constitution, "surely we created man of the best stature."[12] Therefore, it encourages and uplifts in humankind the tendency of good *(ma'ruf)*, virtue *(birr)*, piety *(taqwa)*, and equity (*'adl*). It, at the same time, condemns rebellious and deviant behavior, lewdness *(fahash),* oppression *(taghut)* and exploitation, and insecurity *(fitnah)* of the human *(munkar).*

Each one of the positively and constructively applied ethical categories and attitudes mentioned above calls for a substantial change in behavior pattern of the human being from a hunger and drive-generated stimulus response relationship to an ethically-motivated behavior.

Two broad ethical categories directly related with the core category of equity *('adl)* act as points of reference in cognitive human behavior, in personal matters, and in social, economic, political, legal, cultural, and international matters. These two ethical categories are *halal* (permissible) and *haram* (ethically prohibited and bad). Conceptually, *haram* and *halal* have wider and applied connotations. These terms are often deprived of their wider application when some narrow-minded Muslims confine their application to only dietary rules, dress, or gender relations. The fact of the matter is that these two categories provide guidance in virtually all areas such as economy, law, culture, and political power-sharing, as well as environmental protection and ecology.

## 5. Personal Ethics

The applied and measurable Islamic ethical categories of *halal* (permissible) and *haram* (prohibited) work as the motivating forces for ethically good or bad human behavior. This is reflected in personal matters and in one's social, economic, and other various areas of transactions. Let us take, for example, the case of plantation of poppy used primarily for the production of heroin. In a materialistic and relativistic ethical paradigm, its plantation makes a poor farmer a millionaire overnight. Its plantation and export in a materialistic paradigm is considered no different from the production of other agricultural products, mainly because of the earning of profit from the sale. But the Islamic applied ethics of *halal* and *haram* views the issue differently. Three basic principles are generally raised in this connection.

First, does Islamic ethics allow production of things that distort and pollute human reason (*'aql*) and deprive their users of making independent and rational decisions? The reason given in the Qur'an for the prohibition of intoxicants is precisely that intoxication leads to a state of mind in which a person cannot make an ethically and rationally responsible judgment. Heroin, though not mentioned by name in the Qur'an, is known to cause a more severe state of intoxication and addiction than alcoholic drinks. Consequently, due to the common harmful effect, between the two, heroin shall fall in the category of (*haram*), or prohibited things. Further, the Prophetic *hadith* specifically prohibits, not only consumption of intoxicants, but also their production, storage, sale, and serving. Analogically, plantation of poppy, its trade, storage, transport from one place to another, or any other form of involvement in its trade is considered unethical and unlawful, or *haram*.

The second ethical issue raised by the Muslim scholars relates to the public good. The principle known as *maslahah 'amah* simply means whatever causes *harm* to the public welfare is to be removed. Consequently, production and trade of substances, like heroin, which cause irreparable harm to the coming generations, cannot be allowed. Perhaps this is why the so-called "fundamentalist" regime of the Taliban took a severe action against farming of poppy and during their rule totally destroyed poppy cultivation in Afghanistan. Their policy was unfortunately reversed by the present pro-American government of the Northern Alliance in Afghanistan in the name of "modernism" and "enlightenment."

Third, the ethical principles of *haram* and *halal* make it explicit that anything that causes danger or harm to human life is to be avoided. The Qur'an declares that the unjust killing of one human being is like killing the whole of humankind (*al-Ma'idah* 5:32). Consequently, killing oneself by a known slow poisoning agent, heroin, or similar drugs, cannot be approved.

The Islamic legal maxims declare addiction, as to the extent of becoming a slave to a thing, a violation of the basic doctrine of obedience to Allah. Someone who is addicted to a drug practically becomes a prisoner or slave to it. This violates the basic doctrine of accepting Allah alone as Lord and Authority. Therefore, production and trade of drugs that cause addiction cannot be allowed.

The above three principles have direct relevance to not only the problem of drugs but most of the environmental irregularities and violations. This illustrates how an environmentally friendly and applied ethic, when assuming normative status, influences law, social policy, and development. This normative approach makes the Islamic system directly relevant to the future of humanity. Metaphysical and theological discourses, in general, do not enjoy a central place in the Qur'an. Its applied ethical approach through a rational process leads a critical mind to search for the meaning in creation and in the nature of relationship with the Sustainer and the Protector of humans and the Cosmos.

The critical mind is advised not to confine its research and investigation to one single area. It is welcomed to raise questions about the genesis and legitimacy of ethics as such. Basic issues, such as the normative status of customary practices in a culture, are addressed. Effort is made to understand the viability of a global and universal ethic. Not all of these important questions can be answered in a brief intellectual discourse, such as the present one. However, concerned with the welfare and future of humanity, Islamic sources do highlight some general principles as the basis for an applied ethical and normative paradigm.

## 6. Revelation (*wahi*) as the Transcendent and Objective Source of Islamic Ethics

The Islamic ethical paradigm draws its legitimacy from a transcendent source of knowledge, namely, revelation, or *wahi*. *Wahi*, is regarded in the Qur'an as an objective, not subjective, category. It is not to be confused with inspiration or intuitive knowledge. The Qur'an, in various places, defines *wahi* as knowledge and guidance sent directly by Allah through an external carrier (Archangel Gabriel or Jibril) to His Messengers and Prophets. *Wahi* is not a creation of the prophetic mind. As an objective and external source of knowledge, it is *wahi* that provides universal ethical norms, values, and principles.

Ethical and moral teachings in the Islamic paradigm are not considered a product of a given social and cultural environment. On the other hand, its values, norms, and principles carry objective universal character and applicability. Generally speaking, it is assumed that with the change in space-time moral and ethical practices of people also change. An obvious example, perhaps, is the changing attitude of European and North American society toward homosexuality and lesbianism. What was once considered unnatural, thus illegal and immoral, with the change in society has now been sanctified, legalized, and promoted by some as a "human right." This makes ethics, in the final analysis, situational and subject to social change.

The Islamic ethical system, on the contrary, does not appear to consider same-gender relationships as even an "animal right." For example, Islam does not approve the practice of same-gender relations among monkeys and other animals.

The above example further illustrates that ethical relativism, in which ethics becomes subject to individual or collective human reason and pragmatism, utilitarianism, or intuition, can lead to a normative crisis. What was conventionally regarded unethical, in a changed, modern or post-modern scenario, can become a "human right." If ethical norms and values are drawn from a purely logical and deductive process, social ethics, in a final analysis, will turn out to be subjective, particular, transitory, or subservient to change in society. Ultimacy of human reason or empirical experience or intuition as the origin and source of value system, with all its claims for excellence, remains, nevertheless, particular and confined to a given space-time.

The Islamic paradigm, on the contrary, suggests that universal and global ethical principles, values, and norms can only be given by One whose Knowledge (*'ilm*), Authority (*hukm*), and Intellect (*'aql*) is not finite. This is why, according to the Qur'an, *wahi*, is not a creation of the mind of Prophet Muhammad (peace be upon him). Similarly, the Qur'an states that the revelation sent to Abraham, Moses, and Jesus (peace be upon them) was not a work of their own human mind, but *wahi*, from Allah. A logical consequence of this approach is that global and universal ethical values, originating in the teachings of these true messengers of Allah, carry the potential to become heritage of humankind.

What was declared in the Abrahamic message is consequently reflected in the *shari'ah* delivered by Moses (peace be upon him), and what was declared *haram* or *halal*, in the pristine and original message of Jesus (peace be upon him), is retained in the Qur'anic teachings. The Qur'an refers to divine guidance, Allah's messengers imparted, in a different period and in different parts of the world as valid, legitimate, and universal. A reflection of these global ethical values can be easily traced in the ancient Chinese, African, and South Asian wisdom literature. An epistemic paradigm based on *wahi* from the Creator of humanity thus causes a paradigm shift in the history of humanity and in the moral and social behavior and attitude of human beings.

These absolute ethical norms and principles transcend bondage of color, race, ethnicity, and nationalism, and are under no circumstances subservient to a given space-time.

## 7. Ecological Balance in Civil Society

The Qur'an offers *wahi* as the transcendent and objective source for the development of a balanced and fair attitude in life. It does not consider balance and harmony a monopoly of any particular people. Being universal, these values are to be shared with the whole of humankind and can be translated into the social policy of different civil societies.

The Qur'an also asserts that a unique balance and harmony exists in the universe as an observable sign of the operationalization of the principle of coherence or unity (*tawhid*) at a cosmic level. This ecological balance, interrelatedness, and interdependence are also considered by the Qur'an as signs of willful acceptance of Allah's ultimate authority at a cosmic plane. It says, "Do they seek another way than God's? While unto Him submits whosoever is in the heavens and the earth, willingly or unwillingly and unto Him they will be returned."[13]

The Qur'an maintains that the cosmic ecological balance is due to the operationalization of the principle of coherence and unity (*tawhid*). If, instead of this principle (*tawhid*), some other principles were operating in the cosmos, a logical consequence will be a clash and collision and not coordination, cohesion, and balance in nature.

Translation of the principle of coherence and unity is also expected to take place at a human level, particularly with the development of ethical consciousness among human beings, in terms of *halal*, *haram*, and *mubah*, that is, pure thus permissible; impure therefore prohibited; and initially acceptable. A simple example may illustrate the point. Islamically, it is *halal* to marry through a social and public contract in order to live a family life. However, it is *haram* to have pre-marital privacy or a dating or sexual relationship for the simple reason that such an act is contrary to ethical conduct, social ecology, sanctity of family, and morality. The normative logic behind this is that if free sexual activity is allowed, values of fidelity, ethical accountability, dignity of human relations, and ultimately social balance will be disturbed. Similarly, if people develop a tolerance for unethical social practices, it will lead to public corruption of values.

The culture of *halal* and *haram*, (ethically good or prohibited), thus underscores applications of ethical and moral values in environmental matters. Environmental balance and purity contributes toward realization of social, economic, political, and cultural equilibrium and balance. Social and moral corruption and exploitation, therefore, have a causal relationship with the physical and biological environment.

The concept of a civil society, in secular European thought, moves essentially around individualism and ethical relativism. Consequently, values become relative and subservient to social evolution. In an Islamic framework, on the other hand, civil society is founded on universal and global ethical values. Foremost among them is *'adl*, or justice, equity, and fairness. This primary value is not a product of social development. On the contrary, it provides the foundation and logical basis for development of a just social order. Application of *'adl* in environmental matters implies protection of forests, waterways, and air. Islam primarily wants to construct a social and ethical consciousness in the members of society on the basis of the value of equity and justice. The value of protection and promotion of life also assumes a central place in an ethically conscious society and is reflected in its industrial and technological development. Since *'adl* requires respect for fellow human beings, especially elderly persons and children, development strategies are supposed to make a provision for the welfare of the members of the extended family. These and other social and ethical concerns become an integral part of Islamic environmental ethics. Its concern for environmental purity and balance or ecological consciousness, ultimately helps inculcate in the individual and society an *ethically motivated behavior*.

## 8. The Impact of Economic Imbalance

It appears that our age has become known for extremism, imbalance, and unethical human interventions as a major cause of environmental pollution. Emission of $CO_2$ and other poisonous gases is directly linked with mega-industrial projects and a lust for economic colonialism in the developed and capitalist world. Similarly, water pollution and land erosion are mostly caused by socio-economic policies and the behavior of contemporary human beings.

We all know that the symbols of the contemporary global capitalist order, such as WTO, World Bank, and IMF, in the final analysis, regulate world economy. Their policies have serious impact on the economic development of smaller nations. However, these and other international agencies have totally failed in prevailing over the so-called unipolar imperial power in even getting signed the Kyoto Protocol on toxic emission and global warming. With tall claims about global peace and democracy, U.S. imperialism has not yet improved its record on environmental pollution.

Physical pollution in the environment is directly related to the level of ethical and moral consciousness in the individual and society. The ethical, social, and moral crisis, consequently, is reflected in policies and political strategies such as uncalled for destruction of environment by the indiscriminate and brutal carpet bombing in Afghanistan, in the name of terrorizing "terrorists," and in the name of the realization of "democracy" in Iraq by the U.S. The extent of social and economic disaster, caused by unilateral state terrorism in Afghanistan and Iraq, is yet to be determined. Nevertheless, it has caused irreparable environmental destruction. Wildlife, orchids, birds, and land and water resources were destroyed by the indiscriminate bombing in both areas.

The comparative study of world civilization tells us that whenever a society and a state suffer from double standards of ethics and morality, the ultimate losers are humankind and the environment. The remedy to this crisis, in my view, is not to react in a brutal and uncivilized manner against the so-called mad and crazy "fundamentalists," for no negative attitude or behavior can be corrected by evil and terrorist ways. The positive ethical principle highlighted by the Qur'an in this respect is to "Remember that good deeds (*hasanat*) nullify the bad (evil)."[14] It recommends a proactive, constructive, tolerant, and balanced approach to remove social and environmental evils.

Political power brokers also play an important role in protecting or harming environmental balance and equity. At a conceptual level, the neo-conservatives in the U.S. claim that the capitalist economy and Western democracy and culture are invincible and the ultimate model for the "rest." That humankind has reached the "end of history." Capitalism has emerged as the ultimate order and that all "others" who do not conform with the capitalist value system will inevitably be subject to a collision or a "clash of civilizations." These assertions are questionable. While declaring capitalist political and economic order as the ultimate world order, the capitalist mindset, with all tall claims about democracy, denounces the viability of plurality in society, culture, and environment. It claims the "end of history" has come with ultimate triumph of capitalist values and NWO. Especially after the disintegration of the former Union of Soviet Socialist Republics, the imperialistic mindset has emerged with its self-proclaimed righteousness and obvious arrogance. It does not appear to recognize "others" as cultural and civilizational co-equals. Islam, on the other hand, considers plurality as a hallmark of its civilization. It believes that nature carries an innate diversity and plurality in colors and forms. The geological world, plants, flowers, waterfalls, rivers, lakes, and animals, all stand for variety and plurality. Imposition of only one system, may it be the so-called NWO, or western secular liberal democracy or WTO, is a deviation from plurality reflected in the ecosystems in nature.

The post-globalization challenge we face is, perhaps, no less paramount than the challenge posed by the classical industrial revolution. This new age we have entered is characterized by the concept of a fast flow of capital across the borders, unprecedented developments in information technology, fast means of communication, and the ever-increasing gap between the poor and rich nations of the world.

Globalization has come with deregulation of economy, privatization, and trade liberalization, and with a claim to offer tax reforms, security of property rights, and higher public spending on health and education.

The global economic data, however, show an alarming situation. The world GDP for the year of 2003 was estimated at U.S. $34.5 trillion. Out of this, the U.S., with a GDP of $10.9 trillion, has a share of about 31.6 percent. If we add the GDP of Japan ($4.4 trillion) and Germany ($2.1 trillion), the combined share of the world's three richest countries in global GDP comes to around 50.4 percent. While these three countries have a combined population of hardly eight percent of the world's population, ninety-eight percent of the world's population shares in less than fifty percent of the world's economic resources. This unjust and unethical ever-widening gap between the few rich and the larger part of humanity, the poor, has obvious negative effects on environmental protection.

The economic and social injustice becomes further alarming when we find that about ninety percent of production resources in the developing world including land, stocks, and securities, are owned by the wealthy twenty percent of the population. The remaining ten percent of resources, land, securities, and other similar things, are owned by the remaining eighty percent.

A natural consequence is an unprecedented rise in poverty. The rich nations enjoy the best possible healthcare, education, and production at the cost of the poor and less-developed nations. Poverty of the Third World countries is directly related to this unjust exploitation of the poor by the rich leading to death, starvation, an absence of healthcare, lack of education, pollution, and deforestation.

The developed world upholds only one ultimate value: faster growth rate of capital, maximum profit, and still more profit. If we are serious about environmental health, this survival of the richest has to come to an end. Economy and ecology devoid of social, moral, and educational welfare of the poor and the hungry is no more than a capitalist monster. Not only the environment but human beings in the developing world, consequently, are fast becoming an endangered species.

The pacesetter for electronic governance has not been able to govern its own affairs and demonstrate significant progress in the alleviation of poverty, stopping of deforestation, pollution, decrease in the ozone layer and more importantly in developing a holistic approach in life.

In our view, the survival of the human being and society in the post-globalization age is directly related to an *ethical paradigm shift* and a new ethical world order in which ethical values are integrated into social and economic policy. We need a *global ethic and morality* that look beyond personal gains and seek social good. A post-capitalistic *ethical world order* alone can reduce the chances of the annihilation of the human being.

A pro-active approach, in our view, in order to improve this situation and to avoid the so-called clash of civilizations, is the need of the hour. The hegemony of unipolarism and unilateralism needs to be met, not with contempt, hatred, and violence, but with a *global ethical and moral force*. A change of attitude and behavior or *ethically motivated behavior* alone can save the environment and restore an ecological equation between human beings and the cosmos.

### 9. Building a Global Ethical and Moral Force

This brings us to the final part of our discourse, that is, how to build an ethically motivated behavior, a moral force, and an environmentally friendly attitude. This section will also address the potential roles of academicians and civil society and what steps should be taken, for this purpose, by the elite and morally committed people. These questions call not only for ongoing research and serious investigation, but also for the development of a concrete road map. We believe intellectual honesty demands serious efforts in finding out and sharing tentative solutions for the problems faced by humankind. It is first and foremost intellectual corruption that leads to corruption in state policies. Correct thinking is a prerequisite for a fair and balanced global environment.

The Islamic response to the contemporary ecological challenge can be summarized in a sevenfold strategy that can lead to the construction of a *global ethical force*.

#### A. Development of a Holistic Approach in Life

Observance of dual standards in life cannot lead to ecological balance and environmental security. At the same time, realization of a unified (*tawhidi*) approach does not mean creation of a monotonous and mechanical behavior. It calls for a holistic approach. It implies that honesty, sincerity, and truthfulness, for instance, should be observed, not only in places of worship, but in chambers of commerce, parliaments, law courts, and international forums. If we believe in honesty and truth, our affiliation to one faith or the other, or our carrying a brown, black, yellow, or white skin should not stop us from observing one and same principle in our personal, social, economic, political and international matters. A meaningful environmental sustainability and harmony will only come into existence when we liberate ourselves from wearing different masks at different occasions and immerse ourselves in one single color-adherence to one principle. Referring to this principle, the Qur'an states how believers are to respond to the call for developing a holistic vision in life. "We have taken the coloring of Allah (God); and whose shade is better than God's, Him alone we serve."[15]

It is coherence, not contradiction, duality, or conflict, that becomes the first prerequisite for the development of harmony, balance, fairness, and equity (*'adl*) leading to the building of a holistic personality. A fragmented personality and vision of life can hardly create a cohesive society and environment, while a holistic or unified personality strives to create a meaningful and pluralistic yet unified humanity. It recognizes variety in social existence and in nature yet believes in a common ground that unifies all stake holders in a holistic paradigm. Departure from a holistic *tawhidi* approach, to put it in other words, causes imbalance, inequity, and contradictions at individual as well as social and environmental levels.

### B. Equity (*'adl*), Rights (*huquq*) and Obligations (*farid*)

A natural corollary of the holistic approach, unity and unification in life, is the rejection of an unfair, unjust, and imbalanced attitude and behavior toward human beings and the environment. The term used for the development of a proactive approach for this purpose in the Qur'an is *'adl* (justice, equity, balance, and harmony). Personal, social, and environmental crises appear when we follow extremism and leave the path of balance and harmony. The Qur'an calls an unfair attitude toward human beings and toward the Creator a matter of *zulm* (oppression or injustice). At the same time, it calls for a responsible, balanced, moderate, and equitable behavior or observance of *'adl*. In several places, the Qur'an reminds humankind to not deviate from a balanced behavior and indulge in killing oneself by being unfair to ourselves and to the Creator. "O man, what has made you careless concerning your Nourisher/Lord (*Rabb*), the Bountiful, who created you then fashioned then proportioned you (*fa'adalaka*)."[16] It further calls for an impartial and fair treatment to be observed by all ethically committed persons when they judge between people. "Lo, Allah commands you that you restore trusts to their owners, and if you judge between humankind that you judge justly."[17]

Transcending the legal and dictionary meaning of justice and equity, the Qur'an, while underscoring social dimension of *'adl*, wants humankind to inculcate a dutiful, holistic, and respectful attitude toward parents, wife, children and relatives, in-laws, neighbors, deprived sections of society, and even strangers. Addressing the believers, it says, "Verily Allah has enjoined justice (*'adl*) and doing good, and giving [gifts] to your relatives, and forbidden indecency, impropriety, and oppression. He warns you so that you may remember."[18] This Qur'anic advice has become an integral part of a weekly sermon every Friday, delivered all over the world for the past fifteen centuries in every mosque. It reminds followers of their responsibilities toward other human beings and tries to create environmental and social consciousness.

Equitable, fair, and just behavior, in the Qur'an, also includes a human's attitude toward the resources provided, and favors performed by the Creator for humankind. "Have you not seen how Allah has sent down water from the sky and has caused it to penetrate the earth, as water springs and afterward thereby produces crops?"[19] Elsewhere it says, "And We send the winds saturated [with water] and cause water to descend from the sky and give it to you to drink. It is not you who store it up."[20]

The Qur'an also implies that if humankind does not behave with moderation, balance, and responsibility, the One, who out of His sheer love and compassion for humankind provides pure and unpolluted water through rain and other sources, can also withdraw His favors. Once He withdraws His favors, no scientific intervention can provide humankind with fresh drinking water. It is interesting to note that in the Prophetic supplication for rain, a deep ecological and environmental consciousness is reflected. Specific mention is made of not only human beings but of animals, birds, trees, and land which may die if the Creator does not shower His blessings in the form of rain.

Just and fair behavior herein means the conservation of natural resources, proper and responsible use of water, land, and air resources, preservation of the purity of the environment, and avoidance of pollution as an ethical obligation. Islam's concern for environmental purity is perhaps one reason why smoking is prohibited by Islam.

### C. Promotion of Life

Islamic ethics and law hold a very serious view of the loss of human life due to environmental or other reasons. Except for the compensation for manslaughter or the removal of oppression or treason and corruption in land, no harm can be inflicted to human life ". . . that whosoever kills a human being for other than manslaughter or corruption (*fitnah*) in the land, shall be like he had killed the whole of humankind, and whosoever saves the life of one, it shall be as if he had saved life of all mankind."[21] Elsewhere it says, ". . . and that you slay not the human life which Allah has made sacred, save in the cause of justice (*illa bil haqq*)."[22]

This ethical principle also applies, in general, to animal life. In one of His sayings (*hadith*), the Prophet (peace be upon Him) says, "Whoever killed a sparrow unnecessarily, she will cry on the day of judgment saying, O'Allah your such and such a servant killed me to no avail." This *hadith* shows Islam's concern for promotion of life. It also tells us that the protection of wildlife is an ethical duty; the displacement of birds and animals from their natural habitations is not only unethical, it can also create serious environmental and ecological imbalances.

Human intervention and improper policy planning regarding the environment is a major contributory factor in ecological and social imbalance. One obvious example, in this respect, is the issue of population planning. In some countries, rigid policy of a maximum one child per family has brought to light a serious social imbalance. In such countries, youth are facing a crisis in finding their life partners. In other countries, a low birth rate is the cause of the closure of kindergartens, nurseries, and schools for children due to a lack of young children. The Islamic ethical paradigm offers a middle-of-the-road approach by encouraging engagement in family life once a person is mature enough to make responsible judgments. The Qur'an and the Prophetic *sunnah* declare that it is a social and state responsibility to let a person of marriageable age get married. "And marry such of you who are bachelors."[23] The Prophetic *hadith* (statement) says, "Marriage is my practice whoever rejects it, is not from me." Islamic ethics does not allow suppression of natural social and sexual urges in the name of spirituality. It strongly condemns celibacy and denying conjugal relations in the name of piety. It also disallows intentional destroying of the human gene. Social balance and ecological consciousness thus created persuades a person to respect, promote, and protect life in its various forms.

## D. Inter-faith Harmony

In order to create social synergy, the Islamic ethical paradigm encourages intellectual discourse at an inter- and intra-faith level. A prophetic *hadith* declares that "disagreement in my *ummah* is a blessing." The right to hold a reasoned view and to deliberate and discuss matters in order to reach a consensus is underscored by the Qur'an as well: "And consult with them in conduct of affairs. And when you have resolved, then put your trust in Allah."[24] Consequently, the existence of an ongoing intellectual discourse and the widening of an intellectual horizon and intellectual liberty lead to more than one interpretation of the basic Islamic sources. The Qur'an and the *sunnah* thus become the basis of a living intellectual tradition in the Muslim community (*ummah*).

Acceptance of a variety of views, of religious pluralism, and of the presence of people with more than one cultural tradition, contributes toward a balanced and moderate society. Islam's concern for inter-faith harmony aides in the realization of a human ecology and assists in maintaining a balanced approach toward the environment.

The textual foundations of Islam's concern for pluralism can be easily traced in the general principle that "there is no compulsion in religion (*din*)."[25] It is further elaborated in the statement, "unto you your religion, and unto me my (*din*) way of life."[25] Social and cultural pluralism thus become a distinct feature of an ideal Islamic society and state. These Qur'anic references dispel the common misconception that an Islamic state is theocratic with no room for plurality or intellectual and cultural freedom. The culture of acceptance and recognition of differences develops respect and tolerance for other approaches and views needed for the development of social and cultural harmony.

Religious and cultural plurality realized in the Islamic ethical paradigm is neither exclusivist nor inclusivist, but interactive. It recognizes ground realities, yet believes in the ultimacy of truth. The intellectual argument developed by the Qur'an on plurality is directly related to the cosmic and environmental reality of the existence of pairs and variety in creation. To develop harmony with nature, the human being is also instructed to accept plurality.

## E. Rational Liberty

Ethical judgments in the Islamic framework of thought are based on rational liberty, freedom of the will, and creation of an environment in which reason prevails. This does not mean emotions, feelings, and internal experiences have no legitimacy. It only means rational judgments should prevail over a human's natural emotional attachments and biases. It also implies the frustration in the society, which deters members of the community from maintaining a rational attitude. One obvious example is Islam's total rejection of the drug culture – use and production of intoxicants. A less-than-rational approach can easily lead a person to destroy environmental purity and balance by satisfying his or her personal needs. However, a consumer behavior guided by reason and rational judgment carries chances of being more environmentally friendly. Environmental health can be properly preserved and promoted only when a rational attitude is allowed to prevail in individual and social decisions and policies.

## F. Dignity and Honor

Socio-cultural and environmental health and purity are also closely related to the preservation of dignity and honor of the members of a society. Anonymity is a major enemy of social cohesion and harmony. One obvious reason for juvenile delinquency and teenage violence and crime in our society is the feeling of anonymity, alienation, a generation gap, or sense of rootlessness in the psyche of youth. The marginalization of the family system and the emergence of alternate lifestyles, to the extent of same-gender marriages, is quickly becoming a major cause of the disturbing social equilibrium and value system of our society. Ethically concerned communities must play a proactive role in order to save humanity from the harms caused by anonymity and the uprooting of family values. Social, cultural, economic, and political developments are linked directly with this basic issue. The family provides the foundation stone for society and civilizational development. Deviations, particularly marginalization of family values, weaken the process of civilizational development, disturb social ecology, and make civilizational decline inevitable. If societies, in the name of "human rights," tolerate to allow same-gender marriages or not to live a family life at all, within three generations we may end up in a faceless society. Our culture, ethical values, social system, human relations, or, in brief, the very existence of society, will face a total annihilation.

A cultureless and valueless society can have no concern for the conservation of resources, equitable sharing of water, or land and energy resourcing. Similarly, this society will have no interest in sacrifice or humanitarian help. It will make environmental collapse unavoidable.

### G. Economic Security

Ethical philosophies and systems that do not take into consideration the importance of economic and personal security of the members of society fail to develop a just social order. The Islamic ethical paradigm assigns great importance to economic security in order to create equitable, balanced development and progress in society. It strongly prohibits use of unethical means in economic transactions and encourages transparent economic activity as a moral and social obligation.

The deforesting in Africa, Asia, and South America is directly related to the threat to the economic security and viability in the area. No doubt, moral sermons alone, without economic justice and alleviation of poverty, cannot solve existential and material problems of a people. Nevertheless, if moral values are inculcated and ethical motivation is encouraged and taken as the basis of economic development, environmental disasters can be substantially reduced. This does not minimize the importance of economic human needs. It only suggests a stronger moral and ethical consciousness to guide and direct the existing market behavior.

## 10. Conclusion

With their global relevance and applicability, these seven ethical principles provide a road map for a better ecological and environmental development. In an age of globalization, we need to overcome our biases about others while preserving our own religious and cultural identity. Developing a global outlook, we should plan and engage ourselves in projects for the greater benefits of humanity as one large and extended family.

Today, no country of the world can deny the chain effects that take place whenever a small environmental crisis appears on our planet. The global effects of the recent tragic event, the tsunami, are yet to be properly studied. It is time for ethically concerned persons to develop a *global ethical force* with a commitment to fair, just, and equitable participation of people from different cultural and religious traditions in order to build our world of the future of humanity. The sevenfold global ethical principle can serve as the basis for the future development of a sustainable environment and ecology.

If our environmental policies and our consumer behavior are not founded on sound ethical principles, we will witness frequent brutalities, apathy, and lawlessness as was experienced in New Orleans during Hurricane Katrina in August 2005. Natural disasters, accidents, poverty, and scarcity of supplies provide an acidity test for the strength of the moral and ethical foundation of society and culture. Forging an *ethical and moral force* at a global level alone can help humanity in averting human crisis and calamity.

## NOTES

1. Zafar Afaq Ansari, trans., *The Qur'an, al-Nahl* 16:36 (Leicester, U.K.: The Islamic Foundation, 2006).
2. Ansari, Luqman 31:17.
3. Wayne Ellwood, *The No-Nonsense Guide to Globalization* (Oxford: New Internationalist Publications, 2001), p. 95.
4. *The Guardian*, London, 30 March 2005.
5. *The Qur'an*, al-A'raf 7:57.
6. *Ibid.*, al-An'am 6:100.
7. *Ibid.*, al-Furqan 26:67.
8. *Ibid.*, al-An'am 6:141.
9. *Ibid.*, al-Hijr 15:19.
10. Abdul Hamid Siddiqi, trans., *Mishkat-ul-Masabih*, Vol. 1 (Lahore, Pakistan: Islamic Publications Ltd, 1976), Hadith, No. 427, p. 215.
11. *Ibid.*
12. *The Qur'an*, al-Tin 95:4.
13. *Ibid.*, al-i-'Imran 3:83.
14. *Ibid.*, Hud 11:114.
15. *Ibid.*, al-Baaqarh 2:138.
16. *Ibid.*, al-Infitar 82:7.
17. *Ibid.*, al-Nisa 4:58.
18. *Ibid.*, an-Nahl 16:90.
19. *Ibid.*, al-Zummar 39:21.
20. *Ibid.*, al-Hijr 15:22.
21. *Ibid.*, al-Ma'idah 5:35.
22. *Ibid.*, al-An'am 6:151.
23. *Ibid.*, al-Nur 24:32.
24. *Ibid.*, al-i-Imran 3:159.
25. *Ibid.*, al-Baqarh 2:256.
26. *Ibid.*, al-Kaferun 109:6.

# Seven

# IN SEARCH OF AN ENVIRONMENTAL ETHICS IN EARLY BUDDHISM

Pragati Sahni

## 1. Introduction

The assumption that Buddhist literary sources lend support to an environmental ethics that facilitates the protection of a natural world consisting of different habitats and animals is a common one. Though the possibility of this notion has been upheld by some scholars with great enthusiasm, there exists an equally strong point of view that finds many claims related to the former challenging. In this chapter, I examine some of these opposing claims briefly and as inclusively as possible as also some solutions that are offered that keep both these aspects in view. I also critically reflect upon further ways of establishing more progressive forms of environmental ethics within Buddhism that can then be presented to overcome some of the allegations made at both ends. For this I seek inspiration from the contemporary models of deep ecology and environmental virtue ethics. In order to limit the scope of my enquiry, as I must, for the history of Buddhism is too vast to be looked at simultaneously, I undertake an investigation of some portions of the Pali Canon of Buddhism alone. My deliberations reflect ideas that are present in these texts of early Buddhism and do not include a study of other genres of this Buddhism such as non-canonical literature, art, and so on. I relax this constraint somewhat while reviewing the works of other scholars to get a more extensive picture.

Before I go any further I would like to elucidate very generally what the Pali Canon of early Buddhism signifies. The "Pali Canon" is a collection of literary texts containing the Buddha's teaching that have survived from the very first writings. They are written in the ancient Indian colloquial language of Pali. It must be remembered as a historical fact that though the Pali Canon forms a part of the earliest Buddhism it was transmitted orally for hundreds of years before its texts were recorded and so there exists a substantial time gap between the lifetime of the Buddha and the written version of these texts. The Pali Canon itself has three divisions – he *Sutta Piñaka* containing the discourses of the Buddha; the *Vinaya Piñaka* that includes disciplinary rules for monks; and the slightly later *Abhidhamma Piñaka,* comprising of extensive philosophical and technical discussions. Though I include the study of all three divisions in my study, I concentrate chiefly on the division of discourses as I consider them to represent the method and thought content of early Buddhism with a fervor and distinction that summarizes effectively the spirit of early Buddhist ideas. The Discourses themselves are divided into five parts or *Nikāyas* namely *Dīgha*, *Majjhima*, *Saüyutta*, *Aïguttara*, and *Khuddaka*. The *Khuddaka Nikāya* itself has many sub-parts that include significant texts such as *Dhammapada*, *Theragāthā*, *Jātakas,* and *Sutta-nipāta* amongst many others. My belief is that the views presented hereafter relate to most of the early Buddhist literary corpus quite legitimately. But it must be kept in mind that later forms of Buddhism vary considerably and though similar arguments may be reflected in them, resolutions must be devised by way of reference to them.

## 2. Brief Appraisal of Buddhist Environmentalism

Many scholars believe early Buddhism's message to be one of solidarity and camaraderie with environmental ideas. Martine Batchelor and Kerry Brown have stated in the introduction of their book on Buddhism and ecology that " . . . at its very essence, Buddhism can be described as an ecological religion or a religious ecology."[1] Scholars cite many examples from the literary texts in support of this belief. For instance, the first precept which represents non-violence is repeatedly invoked. The practice of non-violence makes all harming and killing an offence; thus it appears to leave an indelible mark on the treatment of animals and nature in general. Harm to animals and vegetation is viewed by these scholars as objectionable according to the dictates of this precept. According to John Diodo Loori: "The First Grave Precept is *affirm life – do not kill.* What does it mean to kill the environment? It is the worst kind of killing. We are decimating many species. There is no way these life-forms can ever return to the earth."[2] Scholars have noted that the practice of non-violence and compassion are additionally expressed in the Buddha's resistance to the sacrifice of animals. They cite the many occasions depicted in the *Nikāyas* where such sacrifices are severely deplored; alternate sacrifices done with oil, butter and molasses are praised.[3] The practice of non-violence thus serves to reconfirm for them that Buddhism could be regarded as attuned to environmental concerns in a wholly positive way. Several stories in the *Jātaka* collection also convey their displeasure for animal sacrifice and severe punishment for those who indulge in such practices are liberally portrayed. Discussing one such story Christopher Key Chapple says: "This story ridicules the Brahmanical sacrificial process, carrying the message that misguided notions and greed lie at the heart of such behavior . . .. [the story invokes] the Buddhist precept that the life of animals must be protected."[4]

It has been remarked that the detailed list of precepts includes one that forbids injury to seeds and crop. Many scholars draw attention to the major events in the Buddha's life being associated with trees or groves and in this they trace Buddhism's deep connection with and respect for vegetation. Chatsumarn Kabilsingh has pointed out that:

> When we look at the Buddha's pronouncements on water conservation, it is astonishing to see that he actually set down rules forbidding his disciples to contaminate water resources. For example, monks were dissuaded from throwing their waste or leftover food into rivers and lakes, and they were urged to guard the lives all living being abiding there.[5]

Vivid descriptions of natural habitats that became home to wandering monks are evoked by scholars too. Of the *Theragāthā* where many instances appreciating the beauty of nature can be found it has been said that the Buddha and his disciples looked upon nature as a "source of great joy and aesthetic satisfaction" and that they were " . . . sensitive to the beauties and harmony of nature and that their appreciation is colored by spontaneity, simplicity, and a nonsensuous spirituality."[6] Many scholars implicitly believe that Buddhist aesthetics can be used as a resource for conserving the splendor of the natural environment.

Scholars also find a deeper and more foundational solution to the environmental crisis in Buddhism. Speaking of the crisis, Stephen Batchelor states quite compellingly that it exists as the "rather predictable outcome of the kinds of deluded behavior the Buddha described 2,500 years ago. Greed, hatred, and stupidity, the three 'poisons' the Buddha spoke of, have now spilled beyond the confines of the human mind and village politics to poison quite literally the seas, the air and the earth itself."[7] He terms the ecological crisis as a spiritual crisis that demands that persons question the way they live. Batchelor implies that removing these three poisons would positively contribute to environmentalism. Padmasiri de Silva further talks of an environmental psychology when he says: "If we accept the thesis that the pollution of nature and the pollution of the mind are facets of one problem, exploring a viable environmental psychology becomes a significant venture . . . . The psychological roots of ecological disaster and recovery are factors very much related in the Buddhist context to the search for an environmental ethic."[8] Both these writers imply that by cleansing the mind the environment would be purified.

Despite the overwhelming presence of instances such as the above, many problems in the pursuit of an environmental ethics in early Buddhism have been indicated. I have included three of the most obvious and critical here. First, early Buddhism does not incorporate explicit environmental ideas as are employed by contemporary theories. Many essays on Buddhist environmental ethics begin by remarking on this fact. But as seen above (and in what follows) this foundational problem has not discouraged investigation in this area of study – the academic community continues its search due to the widespread unconditional belief that at some level, apparent or obscure, Buddhist literature does address issues of environmental concern. Peter Harvey admits that even though modern ecology and Buddhist principles are not synchronized Buddhism can be envisioned as ecological for it allows for harmony between human beings and nature and for nurture of compassion towards all beings.[9]

The second problem is more taxing – another category of scholars find that a firm adherence to the soteriology of early Buddhism implies moving away from the physical world and its concerns. This becomes problematic for the environmentalist because it negates any argument presented in support of an environmental ethics in Buddhism. Ian Harris, among others, is constantly troubled by this aspect in his musings on Buddhism and ecology especially when he observes that though Buddhists may aspire for the enhancement of the environment, the final end they strive for is quite another. He argues:

> . . . it is dysteleology deeply rooted within Buddhism that is the essential problem . . . . It is not so much that Buddhism has a difficulty in deriving an ought from an is, it is that it faces the more fundamental difficulty of defining an "is" in the first place.[10]

For Harris, the very foundation for building an ethics of the environment is suspect. He indicates that since impermanence, suffering and the relentless cycles of birth and rebirth characterize all existence in this world, the sustenance of an environmental ethics in Buddhism remains challenging.[11]

The third problem pertains to instances where animals and the environment have been mentioned or described in early Buddhist literature. These descriptions are an invaluable requisite for building on an environmental ethics but a large proportion of them are open to different interpretations that may not all contribute to sound environmental theory. Some references, though superficially espousing the cause of nature, on closer examination can be seen as unsympathetic or simply bereft of environmental content. Lambert Schmithausen needs a special mention in this context as his is one of the first detailed studies to express discomfort about such contradictions. Discussing an example that has been made much of in recent environmental research in Buddhism where the practice of universal love or friendship (*mettā*) towards snakes is recommended by the Buddha, Schmithausen remarks that though it happens to be tempting to develop this into an ecological feature, *mettā* was not extended to animals due to "a positive evaluation of tribes and species . . . but rather because these tribes and species are composed of virtually dangerous individuals . . ."[12] that needed to be appeased. And so universal love towards all beings is not for the welfare of the beings themselves but for protection of human beings from the danger these beings pose. Schmithausen is thus unable to find an environmental motivation here. Many other references to nature can be doubted or discredited in a similar manner. The observation that references to some natural entities are inconsistent and vary throughout the texts has also been made as leading to much confusion. In this context Paul Waldau argues that in dealings with animals Buddhism not only typifies a "kinder, first face" but also a "harsher, hierarchically based second face."[13] He notes further the Buddhist tendency to lump together all animals as "of a kind," deprecates animals further.[14] This becomes quite a serious problem for scholars who rely on references to the environment to prove a positive sense of ecology.

Due to such and related concerns many forays in this area have tended to be cautious or have relied on alternative resources within Buddhist philosophy. In a spirit of cautious optimism Schmithausen himself suggests the hermit strand.[15] After grappling with the dominant anthropocentric or human-centered strain in Buddhism, one of the conclusions he draws is that accepting the hermit strand would allow for the existence of anthropocentrism along with ecology. In this he successfully strikes a compromise. His argument is as follows. Hermits are advised to live in natural surroundings in order for them to pursue the path of liberation without distraction. The Buddha himself is depicted as dwelling in forests in pursuit of spirituality. But if there remains no wilderness the recluse would be unable to seek refuge from the trials and tribulations of a social life. Therefore wilderness ought to be preserved (ecology) for the sake of the pursuit of spirituality of human beings (anthropocentrism). Schmithausen proves through such an argument that even though anthropocentrism is a generally criticized concept in environmental literature, the hermit strand has considerable advantages in application to environment and wilderness preservation.

Additionally, though Buddhist doctrines such as dependent co-arising have been tapped as alternative sources by scholars before, I see some of these efforts as also addressing the above outlined chasm between positive and negative views. There is a conscious effort by some scholars to weave an environmental ethics into the very essence of doctrines and thereby avoid condemnation for distorting fundamental Buddhist principles. Alan Sponberg's essay "The Buddhist Conception of an Ecological Self" is one such example where he claims that the Buddhist no-self doctrine can be seen as supportive of environmentalism. "Individual identity" in Buddhism, he believes, is dynamic flow of karma through space and time.[16] Once such a notion of self is absorbed the end result will be a "compassionate, environmental sustaining altruism."[17] David Kalupahana's is another alternative approach that represents the doctrine of dependent co-arising and its application in an environmental ethics.[18] Kalupahana denies the Buddhist approach as absolutistic and implies that the true understanding of dependent co-arising will dissolve the "sharp dichotomies and paired opposites" that arise due to "monstrous solidified conceptual thinking." This dissolution will allow human beings to perceive "a significant relationship between nature and human life."[19] These and many other perspectives are being continually introduced to overcome problems that threaten environmental ethics in early Buddhism.

## 3. Exploring Alternatives

Having summarized quite broadly some supportive and some problematic factors as they exist for scholars trying to understand the status of environment within early Buddhist literary sources in general, I next explore an additional two possibilities for establishing an environmental ethics in early Buddhism. I borrow from two (hugely inspiring though distinctly non-Buddhist) models – namely Deep Ecology and Environmental Virtue Ethics. While I initially begin with comparing these models with some aspects of early Buddhism, I go on to show that despite undeniable similarity early Buddhism holds its own in both instances. The possible ecological undertones present in early Buddhist texts are different from those that characterize these two ways of thinking. But by exploring early Buddhism through the ideas that exist in both these models new ways of understanding early Buddhist environmentalism begin to emerge. I believe that this is the reason such comparisons are beneficial to undertake.

Comparing Buddhism with deep ecology is not a new initiative and their similarities have not gone unnoticed. Deep ecologists themselves have often pointed out the closeness of their philosophy to Buddhism. Scholars including Joanna Macy, David Landis Barnhill, Daniel H. Henning and Deane Curtin have explored themes within Buddhism and deep ecology. It also appears that later forms of Buddhism are found to be more compelling than earlier for such comparisons by writers with some exceptions. I look only to early Buddhism and set out my arguments – some new and others cited previously – briefly. Before beginning my analysis I would like to summarize some important aspects of deep ecology. Though its principles had started to emerge earlier, it was in 1972 that Arne Naess used the term deep ecology as a symbol for a new methodology and movement that wanted to progress beyond surface discourse. Its intention was to "deeply" investigate those concerns that lay beneath the destruction and domination of nature. To deep ecologists this implied examining closely the very ideas and attitudes of persons which came into play in their interactions with the environment. Some went on to claim that values and ethics not deeply rooted in metaphysics almost always led to environmental degradation. Consequently deep ecology aimed towards the awareness and transformation of attitudes and values largely based on materialism and unjustified needs in the belief that this process alone could once more enrich the environment. Naess recognized a shallow ecology by which he implied adopting human-centered and not inclusively thought-out, value-transforming solutions to environmental troubles. In contrast deep ecology suggested the radical transformation of ideological structures the results of which, in the long term, would automatically generate more encompassing and optimal environmental conditions for humans and others alike.

The concept of interrelatedness also appears important in the context of deep ecology. Arne Naess's idea of the perception of identity with greater wholes and later deep ecologist Warwick Fox's notion of transpersonal ecology and process of identification are both based on interrelation.[20] To deep ecologists interrelatedness implies awareness of the links that exist amongst everything in nature such that nature became an integral whole each component or part of which needs or supports another for its existence. Nature therefore has to be viewed in its sum. This view had far-reaching consequences for the development of deep ecological theory. For instance, Naess appeared to believe that the experience of interrelatedness was such that would dissolve the constrictions of isolating individuals from others and through identity with greater wholes "we [would] partake in the creation and maintenance of this whole."[21] Naess' argument has a simple aim: to show that once individual identity gets replaced by a more expansive identity, responsibility would naturally extend in a similar way. Another analogous view was reflected in the works of Aldo Leopold. He suggests that human beings must increase their community to include soil, water, plants and animals and that the right thing is one that upholds the integrity of this community as a whole.[22] Of course extended identity also spelt the end of human centrality and deep ecologists voiced this sentiment by vehemently opposing anthropocentrism or human-centeredness. Instead egalitarianism became a buzz word that stood for the prosperity of each and every entity according to its measure. The state of diversity in nature is appreciated too and deep ecologists also give much importance to ecological intuition as necessary for the realization of all these principles.[23]

Resonances of deep ecology can be glimpsed in early Buddhism. Through adhering to the ethical path the transformation of dispositions and attitudes is envisioned in Buddhism. Early Buddhism's insistence on inculcating ethical values as an essential facet of the Eight-fold Path that then leads to final liberation expresses its conviction in the value of transforming mindsets. Such transformation has an effect on outward actions and dealings as well. Action or *kamma* in Buddhism pertains to thought, speech and action after all.[24] Thus deep ecology's methodology of identifying harmful ideology and re-orienting the thought process had already been identified and traced in early Buddhism many centuries prior. Another aspect of the Eight-fold Path is also reflected in deep ecology, i.e., wisdom or intuition. Ecological intuition in deep ecology emphasizes on "no firm ontological divide in the field of existence."[25] Though early Buddhist texts do not necessarily speak of a similar ecological intuition, the development of a universally applicable wisdom (*paññā*) is emphasized. Deep reflection before a deed is enacted is also encouraged. Additionally deep ecological wisdom suggests incorporating a lifestyle that remains "simple in means and rich in ends" implying the use of fewer resources for greater satisfaction.[26] In persuading its monks and nuns to be satisfied with few necessities and to dwell in natural surrounding (as suggested in the hermit strand mentioned above) the religion finds more in common with deep ecology. Deep ecology's strong anti-stance on human supremacy can be compared with early Buddhism's praise of humility (*hirī*) and attack on arrogance (*ahaṅkāra*). Both humility and arrogance are often indicated in the texts and are used to deal with the sense of self (both values will be discussed in some detail in the next section).

There are however grave differences in the understanding of interrelation between deep ecology and early Buddhism. The notion of interdependence in Buddhism does not necessarily extend to interrelation. It is true that early Buddhism recognizes *paṅiccasamuppāda* or dependent co-arising, that is, that consequences result from precise causes, as a universal truth.[27] But nowhere do the Buddhists go beyond affirming a theory of causation. Many scholars including Schmithausen therefore object to such parallels in general.[28] Early Buddhism also has a completely different approach to questions of identity and I don't believe that the notion of an expansive self can be found within its literature. Additionally discussions on the nature of a metaphysical self are complicated and often end in denial or silence. With such a foundation it becomes exceedingly difficult to speak of "identifying with greater wholes." There also exists a definite impression that early Buddhism does not reject human-centeredness as vehemently as deep ecology. Since it exists as a philosophy focused on ultimate spiritual emancipation achievable only in the human realm, human beings have a special value. Being repositories of special value could imply that human needs must be met more urgently and this may lead to collision with the principles of deep ecology.

Not being able to identify with alacrity early Buddhism and deep ecology does not imply failure of the initial objective. From the comparison emerge some crucial points that allow early Buddhism to express its very own ecological principles. I discuss some instances here. Though *paṅiccasamuppāda* does not support interrelatedness of a deep-ecological kind it does appear to support the universality of causation operating in all entities and processes in the universe irrespective. In giving a formula that spares no one Buddhism illustrates a unifying factor *and* questions the self-importance that human beings have given themselves, based on the belief that they have special, distinctive value. Buddhism's uniqueness is further amplified when it represents both human-centrality and non-arrogance with equal gusto. Its so-called anthropocentrism runs parallel with compassion and loving-kindness making selfishness or speciesism impossible. This can be evinced from many instances in the texts that ask practitioners to look out for their interests as also for those of others.[29] In summary it is obvious from the deep ecology model that its focus is in the direction of foundational issues – it questions the superior value of human beings, it aspires for the flourishing of human as well as non-human life through ideological change, and it encourages the expansion of intuition and awareness – and in this it provides a method or a process through which the environmental crisis could be challenged. In my account I have shown that the early Buddhist outlook is compatible with some deep ecological principles and at the same time it challenges some of the accepted notions of this system of thought. However more notably the dialogue has made new ideas emerge from Buddhism. Aspects of environmentalism from early Buddhism that may not have been known and or made explicit have been revealed. A semblance of new environmental morality begins to materialize through a respect for the difference with deep ecology.

## 4. Environmental Virtue Ethics in Early Buddhism

The above deliberations on the principles of deep ecology being expressed in early Buddhism also inspire an investigation into the presence of an environmental virtue ethics in this Buddhism. This form of environmental ethics has been developed and analyzed by scholars including Thomas E. Hill Jr. and Louke van Wensveen, to name a few. To them the character of the agent is of immense relevance and must be taken as a central factor of ethical discussion. Significantly environmental virtue ethicist Geoffrey Frasz asks the questions, "What sort of person would wantonly destroy natural entities?" or "What sort of personal qualities are needed for the humane treatment of nonhuman creatures?"[30] Pondering over such questions leads these ethicists quite naturally to study the virtues and ideals that would influence the state of the natural world. Virtues or then again lists of virtues are examined by them with this aim in mind. Hill for instance spoke widely of the virtues of humility and altruism. In what is widely regarded as one of the first writings on environmental virtue ethics he argues in favor of the "ideals of human excellence" for preserving natural environments."[31] Hill's argument hinges on the aspect that environmental attitudes are linked with general virtues and a lack in general virtues would signal paucity in the former as well. In giving more form and substance to an emerging environmental virtue ethics Wensveen talks among other things of an ecological virtue language such as "has a distinct vocabulary and operates according to a discrete set of conventions" and ". . . with this new virtue language we may be given not just another chance but a *good* chance at achieving moral breakthroughs, at finding fresh ways of looking at problems and dilemmas that are already getting stale . . . ."[32]

In all practical senses tracing the existence of a virtue ethics in early Buddhism is almost instinctive. There exists an overwhelming concern about developing the character of the agent in the early Buddhist scriptures under consideration (even though the presence of deontological and consequential strands cannot be denied). This has not gone unnoticed by scholars. For instance Padmasiri de Silva acknowledges, " . . . Buddhist ethics brings into play a wide variety of virtues for the building up of human character . . . the

Buddha gave a central place to . . . compassion, generosity and gratitude . . . ."[33] Scouring through the texts reveals without a doubt that the presence of virtues is a central thesis and almost all texts gravitate towards a training that would ensure their acceptance and development. Interesting to note is that not just the presence of virtues tilts the scales in favor of a virtue ethics but also the early Buddhist comprehension of other concepts such as *kamma*. The doctrine of *kamma* that can be looked upon as the inevitable nexus between action and its consequence stresses on the character of the agent. According to it moral character will lead to wholesome actions and eventually good consequences and an immoral character to the opposite. Then again when the scriptures speak of the path that would lead to the highest transcendental goal in early Buddhism – liberation or *nibbāna* – they not only suggest the importance of meditation and wisdom but also of *sãla* or morality. The three (comprising the Eight-fold Path) must be followed all together for attaining the goal as also for removing greed (*rāga*), hatred (*dosa*), and delusion (*moha*) that are considered as the biggest impediments to *nibbāna*. So both the quest for liberation and *kamma* surreptitiously incorporate moral training – thereby character development – for which adopting the virtues is of utmost importance.

Sympathy (*anukampā*), compassion (*karunā*), universal love (*mettā*), non-injury (*ahiüsā*), non-covetousness (*anabhijjhā*), and generosity (*dāna*) are virtues that are commonly mentioned in early Buddhism among others. Additionally are mentioned progressive lists of precepts (five, eight, and ten) that contribute to the building of a high moral character. The first five precepts include abandoning the taking of life (*pàõàtipàta veràmaõā*), abandoning the taking of what is not given (*adinnādānā verāmaõī*), abandoning unchaste behavior (*kāmesu-micchācāra verāmaõī*), abandoning false speech (*musāvāda verāmaõī*), and refraining from intoxication of any kind (*surā-meraya-majja pamādaññhānā verāmaõī*).[34] Equal importance is given to the elimination of vices or defilements such as covetousness (*Abhijjhā*), greed (*giddhilobha*), ill-will (*āghāto*), avarice (*macchariya*), and egoism (*ahaükāro*).[35] The list of *kusala* and *akusala cetasikas* given in the *Abhidhamma* is also reminiscent of virtues and vices as listed above.[36] However even though the path of virtues remains reasonably established problems in ascertaining an environmental virtue ethics remain. The above listing of the precepts, virtues, and vices does not have an obvious environmental agenda. For that matter nor do *kamma* and the Eight-fold Path; both do not appear to have much to do with the protection of the natural world. Other questions are more basic and pertain to matters such as whether the practice of ethics can be sustained at higher spiritual levels. A denial would imply abandoning an environmental ethics altogether. This concern is immediately quelled by the belief of many scholars who declare that ethics in Buddhism cannot be neglected by the practitioner at any level. James Whitehill is one such scholar who believes that in Buddhism there exists always a need for morality or ethics. He writes: "The path of Buddhism does not dissolve character (which is different from ego and personality). It awakens and illuminates moral character . . . ."[37]

Dealing successfully with such difficulties as the one above does not absolve from the primary question of whether an environmental virtue ethics can be justified within early Buddhist scriptures whose concern is attaining liberation and not an optimal state of environment. All theories and concepts these scriptures employ are designed to meet this one goal. As mentioned earlier, the embracing of virtues, the practice of the Eight-fold Path, and the pursuit of good *kamma* move in the general direction of spiritual freedom. This fundamental truth does not alter and cannot be denied of the scriptures in any way. But that the virtues, being what they are, unwittingly do contribute to an environmental ethics as well is an idea worth looking into. I have vociferously argued in support of this position elsewhere (given below is a skeletal outline of my more detailed argument).[38] Here I discuss its possibilities briefly and some issues about methodology including what can be gained through such a comparison.

The best way forward appears to be a detailed analysis of the different virtues in order to determine their ecological significance. In keeping with this I examine two as exemplars – *ahiüsā* and *hirī*. Translated generally as non-violence and humility respectively the importance of both the virtues has been constantly stressed by many environmentalists. *Ahiüsā* as a virtue is central in early Buddhism. The first precept speaks of *ahiüsā* (that I also understand more broadly as "respect for life") and can be taken to imply non-aggression and the use of peaceful means. Not only is the performance of this virtue praiseworthy in others but it is one that is personally practiced by the Buddha himself. Statements such as the following occur often:

> Having forsaken the destroying of life, the ascetic Gotama abstains from destroying life (*Pàõàtipàtaü pahàya pàõàtipàtà pañivirato Samaõo Gotamo*), without stick, without weapon (*nihita-daõóo nihita-sattho*), humble, with kindness, he lives with compassion for the welfare of all living beings (*lajjã dayàpanno sabb- pàõa-bhåta-hitànukampã viharatã ti*).[39]

In the *Dhammapada* is stated that each being embraces dearly its own life and thoughts of pain cause agitation in all; for this reason harming or causing pain to others must be avoided (*sabbe tasanti daõóassa, sabbesaü jīvita piyaü, attānaü upamaü katvā, na haneyya na ghātaye*).[40] This appears as a Buddhist version of the golden rule. Virtues of non-violence and non-hurting are exhorted in the *Theragāthā* too.[41] Right action (*sammà kammanta*) and right livelihood (*sammà àjãva*) that are embodied in the Eight-fold Path also support non-violence.[42] These target a change in lifestyle whereby the personal and professional life of the agent is so structured that harm to other beings is out of question. The Buddha apparently disapproves of dealings in weapons and meat production – indicating a dismissal of those forms of making a living that entail violence and hostility.[43] This point is further highlighted when statements are made in the scriptures about some professions such as those of hunters and butchers as taboo or prohibited.[44] The ban on animal sacrifice also goes to show that violence against other beings was intolerable and severely criticized.[45] Non-injury to seeds and plant life was also suggested in the detailed precepts and the *Vinaya*.[46] Compassion (*karunā*) is encouraged and this gives more credence to all aspects of non-violence. But as noted by scholars and as mentioned earlier in this chapter, all these approaches to non-violence (for instance towards plants) have little to do with environmental ethics.[47] They are more about the agent's spiritual escalation or other matters. This is a common criticism levied against scholars who rely upon such arguments to prove early Buddhism's environmental character.

To focus on this criticism alone however would be missing the point I am making here. Respect for life as a virtue is more encompassing in early Buddhism and not limited to the above instances typifying *ahiüsā* from the scriptures. It indicates adopting this virtue in thought and deed. Once adopted and internalized the virtue is bound to have an effect on all aspects of life including the treatment of the environment. But it becomes especially significant for the environment as it can form the basis of the protection of plant and animal life. And its special value lies in its ability to do this without interfering with early Buddhism's ultimate goal of liberation.

Another often mentioned environmental virtue is humility. This virtue has already been mentioned with reference to deep ecology. Humility is crucial to an environmental virtue ethics because it counters feelings of superiority that human beings may feel. Crucially the arrogance and lack of humility of the human species is often blamed for depleting natural resources because it implies that human needs and desires must be met without cognizance of the life and needs of non-humans. Optimistically speaking once adopted this virtue can neutralize arrogance or feelings of superiority thereby allowing space for regarding of the natural world. Though early Buddhist scriptures do not speak of humility in this exact sense it retains importance in a potential environmental virtue ethics in the religion due to the possible effects of its practice. The Buddha is referred to as exercising *Lajjin* and this can be translated as modesty or conscientiousness.[48] The Pali term *hirī* sometimes taken to mean a sense of shame is also understood by scholars as humility or modesty.[49] The latter exists as a central virtue in early Buddhism and is mentioned in conjunction with spiritual practice. For instance the *Saüyutta Nikāya* states that higher levels of practice will not be attainable were this virtue neglected.[50] On the contrary, the agent's practice will see a decline. *Hirã* is also praised in the *Jàtakas*. The *Devadhamma-jàtaka* tells the story of a demon that does not have humility. Not only does the *bodhisattva* (here the Buddha in a previous incarnation) bring him to the path of humility but also expounds the benefits of following the path of virtues. The story ends with the demon becoming virtuous.[51]

A visible attack on arrogance or *ahaükāra* corroborates the above described essence of the virtue. The *Sutta-Nipāta* states at one place that a monk who has overcome arrogance as completely as a flood of water wipes out a frail bridge of reeds has come closer to attaining liberation.[52] An interesting nature related instance where arrogance is mentioned occurs in the *Aggañña* Discourse of the *Dãgha Nikāya*. This discourse is generally referred to as containing the Buddhist version of the creation of the world. The arrogance of human beings here is shown to lead to a colossal transformation in the natural environment (*tesaü vaõõātimāna-paccayā mānātimāna-jātikānaü rasa-pañhavī antaradhāyi*) – not necessarily in an affirmative sense.[53] Other doctrines and ideas within early Buddhism hint at nurturing humility. For instance the doctrine of no-self or *anattāvāda* does not allow the agent to resort to egoism. There exists no permanent self, the doctrine proclaims, making the basis of individualism and preoccupation with the personal self dissolve. The self has five factors or *khandhas*, all impermanent, whose partnership ends on the death of the person. The agent arrives at perfect knowledge once this truth is comprehended.[54] Under such circumstances it becomes almost impossible to sustain egoism and its resultant arrogance. It must be remembered here, as in the case of *ahiüsā*, that the virtue of *hirī* was never expressly suggested for the sake of environmental practices. But it corresponds with an environmental virtue (as has been mentioned by ethicists such as Thomas E. Hill Jr.) and there exists no reason for it not to have a similar environmental effect amongst its other effects. It becomes an environmental virtue by virtue of its description.

Having examined these two virtues generally it would be advantageous to anticipate some objections that continue to persist. Scholars have discussed similar issues in virtue and environmental virtue theory and these automatically come to be levied against an early Buddhist adaptation as well. First, due to the absence of a proper definition of an environmental virtue in early Buddhism what can or cannot be classified as one becomes obscure. Added to this is the objection that there is lack of guidelines about how a choice is to be made between contrasting environmental virtues when each is perfectly legitimate. There are very few arguments and examples where such doubts appear to be settled in early Buddhist texts. The validity of the above two objections cannot be questioned but solutions could be worked out to address them. It is true that obscurity with regard to the definition of environmental virtues is a foundational problem in early Buddhism. This is further complicated by the difficulty faced in defining what a virtue is even though the Buddhist texts are flooded with instances of virtue-like traits that scholars have often named virtues. I too have presumed the existence of virtues in Buddhism due to the overwhelming evidence I drew attention to earlier in this section. The need to adopt a process of looking at each trait individually to determine its virtuous nature is indicated. For an environmentalist this process must extend also to determining the trait's possible environmental character. This would be a time consuming but ultimately rewarding process which has been amply demonstrated in the case of respect for life and humility above. The nurturing of both of these qualities has no given environmental agenda but, I believe, can even then be engaged positively in an environmental quest.

The second objection surfaces when virtues conflict. An environmental virtue ethics constantly encounters the dilemma of choosing one virtue over another. Buddhism says little on the relative value of its so-called virtues but is not completely silent either as can be gathered from an interesting example in Buddhist literature where such a choice appears to have been made. I have discussed this example elsewhere and present it below in a few words. The *Jīvaka* Discourse of the *Majjhima Nikāya* deals with the rather controversial subject of meat-eating and becomes of value here because it reveals the grounds on which one of two virtues can be decided upon. On the one hand, the Discourse appears to support the eating of meat as alms by monks to cultivate non-attachment and on the other there exists the extreme early Buddhist reaction to violence, malice, cruelty, and the professions of butchering and hunting. Thus are presented to the reader *ahiüsā* and non-attachment. Since meat-eating is permitted the Discourse inadvertently chooses the second and renounces the first. However it offers grounds for doing so. An elaborate explanation of when and why meat can be eaten is given: if it is seen, heard or suspected however that the animal was killed for the sake of the monk (*tãhi kho ahaü Jãvaka ñhànehi maüsaü aparibhogan- ti vadàmi: diññhaü sutaü parisaïkitaü)* then it becomes an act deserving retribution.[55] Thus meat-eating is qualified. The Discourse in this way hints that the question of picking this or that never arises. It implies that a person must rise above the duality of choices. The scholar Rupert Gethin expresses this sentiment strikingly when he says that "Ultimately Buddhism teaches that the nature of good conduct is subtle and complex – so complex that it precisely cannot be solved by reference to precepts and rules of conduct."[56] He adds that the basis therefore must be "a path of training that ends in rooting out greed, aversion and delusion." He thus brings out the importance of rising above dualities in early Buddhist morality.

Adopting a virtue approach is a good way of challenging some of the above listed problems of Buddhist environmental theory, especially those highlighted by Schmithausen and Harris. The practice of virtues becomes so central that it draws attention away from specific examples pointed out by Schmithausen that appear to be indifferent to the plight of nature. Such examples may therefore hold true within their limited contexts alone but not on a larger scale such that they become models that overwhelm the virtues. Similarly since the practice of virtues not only promotes the individuals own spiritual growth but also inadvertently supports the development of social interactions that consider the welfare of others, the individual has considerably more reason than indicated by Harris to care for nature and environment. The above discussion also reveals that Buddhism's environmental virtue ethics is its very own for it comes inherently merged with a spiritual agenda and can never be treated as a stand alone theory. Only once Buddhism is looked upon in this new light will the uniqueness of its environmental ethics become more significant.

## 5. Conclusion

In conclusion it can be stated that while deep ecology leads researchers to find the more deep-seated causes of the environmental crisis, environmental virtue ethics leads them to posit moral behavior as the central aspect of environmental discourse. Neither is evaluating the principles of deep ecology within early Buddhist scriptures nor investigating the virtues new areas of exploration for environmental ethicists working in Buddhism. Both have been mentioned countless times and approached in many different ways.[57] However it becomes crucial to give these approaches some legitimacy and to concretize their value in showing a new direction to proceed within early Buddhist texts as has been my aim throughout. But this does not overrule some inescapable problems that this method of arriving at Buddhist environmentalism can suffer from. First, it may be the case that comparisons with non-Buddhist models may not only widen the scope but may quite as easily limit possibilities within early Buddhist literature itself. Such comparisons can restrict thinking patterns within certain parameters such that early Buddhism's more exceptional notions get obscured. Other deeper problems continue to persist as well. One is whether deep ecology and environment virtue ethics themselves have the force to deal with the environmental crisis exist as well. It is likely that some of their concerns get passed on to early Buddhism even if the Buddhist theory shapes up with originality, in its distinct way. Would early Buddhism have the resources to deal with such concerns? The need to inquire in detail into what underlies the environmental crisis and to comprehend intimately the extent to which virtues may have an effect on contemporary ecological issues in the early Buddhist context must be engaged with during another occasion. Many questions eventually remain. At the same time many possibilities emerge that must be given serious and detailed thought for taking forward the mantle of establishing a viable Buddhist environmental theory.

## NOTES

1. Martine Batchelor and Kerry Brown, ed., *Buddhism and Ecology* (London, UK: Cassell Publishers Limited, 1992), p. viii.

2. John Diado Loori, "The Precepts and the Environment," *Buddhism and Ecology: The Interconnection of Dharma and Deeds*, eds. Mary Evelyn Tucker and Duncan Ryåken Williams (Cambridge, Mass.: Harvard University Press, 1997), p. 179.

3. *Dīgha Nikāya*, Vol. I/141 (London: Pali Text Society).

4. Christopher Key Chapple, "Animals and Environment in the Buddhist Birth Stories," *Buddhism and Ecology: The Interconnection of Dharma and Deeds*, eds. Mary Evelyn Tucker and Duncan Ryūken Williams (Cambridge, Mass.: Cambridge University Press, 1997), p. 140.

5. Chatsumarn Kabilsingh, "Early Buddhist Views on Nature," *Dharma Gaia*, ed. Alan Hunt Badiner (Berkeley, Cal.: Parallax Press, 1990), p. 11.

6. Lily de Silva, "Early Buddhist Attitudes towards Nature," *Dharma Rain: Sources of Buddhist Environmentalism*, eds. Stephanie Kaza and Kenneth Kraft (Boston, Mass.: Shambhala, 2000), pp. 101–102.

7. Stephen Batchelor, "The Sands of the Ganges," *Buddhism and Ecology*, eds. Martine Batchelor and Kerry Brown (London, UK: Cassell Publishers Limited, 1992), p. 33.

8. Padmasiri de Silva, "Buddhist Environmental Ethics," *Dharma Gaia*, ed. Alan Hunt Badiner (Berkeley, Cal.: Parallax Press, 1990), pp. 16–18.

9. Peter Harvey, *An Introduction to Buddhist Ethics: Foundations, Values, and Issues* (Cambridge, UK: Cambridge University Press, 2000), pp. 185–186.

10. Ian Harris, "Causation and 'Telos': The Problem of Buddhist Environmental Ethics," *Journal of Buddhist Ethics*, 1 (1994), p. 54. (Journal on-line, accessed 7 Oct 1999, available from http://jbe.gold.ac.uk)

11. *Ibid.*, pp. 15–16.

12. Lambert Schmithausen, "The Early Buddhist Tradition and Ecological Ethics," *Journal of Buddhist Ethics*, 4 (1997), pp. 18–19. (Journal On-line, accessed 7 October 1999, available from http://jbe.gold.ac.uk)

13. Paul Waldau, "Buddhism and Animal Rights," *Contemporary Buddhist Ethics*, ed. Damien Keown (Richmond, Surrey: Curzon Press, 2000), p. 105.

14. *Ibid.*, pp. 94–98.

15. Schmithausen, "The Early Buddhist Tradition and Ecological Ethics," pp. 26–28. Lambert Schmithausen, *Buddhism and Nature* (Studia Philologica Buddhica Occasional Paper Series VII, Tokyo: The International Institute for Buddhist Studies, 1991), pp. 18n.

16. Alan Sponberg, "The Buddhist Conception of an Ecological Self," *The Sound of Liberating Truth: Buddhist-Christian: In Honour of Frederick J. Streng*, eds. Sallie B. King and Paul O. Ingram (Richmond, Surrey: Curzon Press, 1999), pp. 116–118.

17. *Ibid.*, p. 118.

18. David J. Kalupahana, "Toward a Middle Path of Survival," *Nature in Asian Traditions of Thought: Essays in Environmental Philosophy*, eds. J. Baird Callicott and Roger T. Ames (Delhi: Sri Satguru Publications, reprint, 1991), pp. 247□256.

19. *Ibid.*, p. 252.

20. Warwick Fox, *Toward a Transpersonal Ecology: Developing New Foundations for Environmentalism* (Boston, London: Shambala, 1990).

21. Arne Naess, *Economy, Community and Lifestyle: Outline of an Ecosophy*, trans. David Rothenberg (Cambridge: Cambridge University Press, 1989).

22. Aldo Leoplold, "The Land Ethic," *A Sand County Almanac, and Sketches Here and There* (New York: Oxford University Press, reprint, 1987), p. 204n.

23. Warwick Fox, "Deep Ecology: A New Philosophy of Our Time?" *Environmental Ethics: An Anthology*, eds. Andrew Light and Holmes Rolston III (Malden, Oxford: Blackwell Publishing Ltd., 2003), pp. 252□261.

24. *Aïguttara Nikāya*, Vol. III/415 (London: Pali Text Society).

25. Fox, "Deep Ecology," p. 255.

26. Arne Naess, "The Deep Ecological Movement: Some Philosophical Aspects," *Environmental Ethics: An Anthology*, eds. Andrew Light and Holmes Rolston III (Malden, Oxford: Blackwell Publishing Ltd., 2003), p. 273

27. *Majjhima Nikāya*, Vol. III/63 (London: Pali Text Society); *Saüyutta Nikāya*, Vol. II/25 (London: Pali Text Society).

28. Lambert Smithausen, "The Early Buddhist Tradition and Ecological Ethics."; Ian Harris, "Buddhist Environmental Ethics and Detraditionalization: The Case of Eco-Buddhism," *Religion* 25 (1995), pp.202f.

29. *Saüyutta Nikāya*, Vol. V/169; *Aïguttara Nikāya*, Vol. II/ 95.

30. Geoffrey B. Frasz, "Environmental Virtue Ethics," *Environmental Ethics*, 15, no. 3 (1993), pp. 259□260.

31. Thomas E. Hill Jr, "Ideals of Human Excellence and Preserving Natural Environments," *Environmental Ethics: What Really Matters, What Really Works*, eds. David Schmidtz and Elizabeth Willott (New York, Oxford: Oxford University Press, reprint, 2002).

32. Louke van Wenveen, "The Emergence of Ecological Virtue Language," *Environmental Virtue Ethics*, eds. Ronald Sandler and Philip Cafaro (Lanham, Boulder: Rowman and Littlefield Publishers, 2005), p. 21 and p. 27.

33. Padmasiri de Silva, "Buddhist Ethics," *A Companion to Ethics*, ed. Peter Singer (Oxford: Blackwell Reference, 1993), p. 64.

34. *Dãgha Nikāya*, Vol. I/4.

35. *Majjhima Nikāya*, Vol. III/50 (London: Pali Text Society); *Majjhima Nikāya*, Vol. I/360; *Dãgha Nikāya*, Vol. I/3; *Majjhima Nikāya*, Vol. I/281; and *Majjhima Nikāya*, Vol. III/18, respectively.

36. Lama Govinda, *Psychological Attitude of Early Buddhist Philosophy* (Delhi: Motilal Banarsidass Publishers Private Limited, reprint, 1991), pp. 115□125.

37. James Whitehill, "Buddhism and the Virtues," *Contemporary Buddhist Ethics*, ed. Damien Keown (Richmond, Surrey: Curzon Press, 2000), p. 23.

38. Pragati Sahni, *Environmental Ethics in Buddhism: A Virtues Approach* (Abingdon, Oxon: Routledge, 2008).

39. *Dăgha Nikāya*, Vol. I/4.

40. *Dhammapada*, trans., Narada Thera (Taiwan: The Corporate Body of the Buddha Educational Foundation, 1993), Verse 130.

41. *Theragāthā*, 602□603 (London: Pali Text Society).

42. *Dăgha Nikāya*, Vol. II/312.

43. *Aïguttara Nikāya*, Vol. III/208.

44. *Majjhima Nikāya*, Vol. I/343.

45. *Dăgha Nikāya*, Vol. I/141.

46. *Dăgha Nikāya*, Vol. I/5, *Vinaya*, Vol. III/155 (London: Pali Text Society).

47. Lambert Schmithausen, *The Problem of Sentience of Plants in Earliest Buddhism* (Studia Philologica Buddhica Monograph Series, VI, Tokyo: The International Institute for Buddhist Studies, 1991).

48. *Dăgha Nikāya*, Vol. I/4.

49. Damien Keown, *The Nature of Buddhist Ethics* (Basingstoke, Hampshire: Palgrave, 2001).

50. *Saüyutta Nikāya*, Vol. II/206.

51. *The Jàtaka or Stories of the Buddha's Former Births*, trans., E. B. Cowell (Delhi: Motilal Banarsidass, reprint, 1990), Vol. I/126□133.

52. *Sutta-Nipāta, A Collection of Discourses*, trans., V. Fausboll, ed. F. Max Muller (Delhi: Motilal Banarsidass Publishers Private Limited, reprint, 2004), Verse 4.

53. *Dăgha Nikāya*, Vol. III/86.

54. *Saüyutta Nikāya*, Vol. III/44□45.

55. *Majjhima Nikāya*, Vol. I/369.

56. Rupert Gethin, *The Foundations of Buddhism* (Oxford: Oxford University Press, 1998), p. 173.

57. Sponberg, "The Buddhist Conception of an Ecological Self"; David J. Kalupahana, *Ethics in Early Buddhism* (Honolulu: University of Hawaii Press, 1995); David E. Cooper and Simon P. James, *Buddhism, Virtue and Environment* (Aldershot: Ashgate, 2005).

# Eight

# ECOSYSTEM SUSTAINABILITY: A DAOIST PERSPECTIVE

Jonathan Chan

Daoism does not appear in its original form as a philosophy advocating some environmental ethic. It does not focus on concern about the problem of how to preserve conditions that will make life worth living in the future, for instance. Nor does it focus on any concern about environmental problems which catch much of our attention nowadays: global warming, air pollution, endangered species extinction, toxification of food chains due to the use of toxic substance by industry and agriculture and so on. Still, one may ask whether Daoism contains some sort of environmental ethic or whether it provides a justification for some form of environmental ethics or not? Some Daoist scholars cast doubt on such a Daoist environmental ethics project. Paul R. Goldin, for instance, argues that Daoism is not environmentalism. There are two lines of reasoning, according to Goldin, which suggest that Daoism provides no environmental ethics at all. The first line of argument is that the actual practices toward nature in Chinese history are anti-environmental. The second line of argument is that Daoism does not advocate ethical notions or precepts out of a conviction that other attitudes toward nature might threaten the environment. The point of this argument is that "if Daoist philosophers did not assert explicitly that their arguments might be applied to concerns of ecology or environmentalism, then promoting their work on the grounds that it might help us manage our environment represents a falsification of their ideas and a self-serving appropriation of the past for purposes grounded in the present."[1] In response, I will make the following rejoinders. First, because of the actual practices toward nature in Chinese history are anti-environmental, it does not follow that Daoism provides no environmental ethics at all. It is because there can be discrepancy between thought and practice. Also, Daoism should not be the only school of thought that is responsible for people's practice in Chinese history. Second, while Daoists did not have environmental concern that people nowadays have, it does not mean that Daoism does not have ethical precepts or principles that favor environmentally sustainable practices. For instance, because scientists are not aware of the technological application of their scientific theory does not prevent the theory from having that application.

The above rejoinders do not show that Daoism has ethical precepts or principles that support environmentalism. However, there are an increasing number of Daoist scholars who are optimistic about developing an environmental ethics from the Daoist perspective. There have been studies on the application of the Daoist perspective to environmental debates since late 1980s. In drawing on the rich intellectual resources of Daoism, many of these studies support the idea that Daoism does have ethical resources to meet the challenge of developing a sound environmental ethic.[2] Cheng Chung-ying, for instance, argued that *Dao* and *qi*, two important metaphysical notions of Daoism, can provide a metaphysical foundation for the five axioms of environmental ethics.[3] Some argue from the aesthetic point of view that Daoism's aesthetic approach to nature provides environmental philosophy with an important conceptual foundation.[4] Some others argue that Daoism and Ecofeminism have strong affinities which "offer possibilities for developing the relevance of both philosophies to bear upon a theoretical understanding of how we can live in a respectful and sustainable relationship with our natural environment."[5] Another Daoist scholar, Karyn L. Lai argues that "[t]he concepts *dao* and *de* in the *Daodejing* may be evoked to support a distinctive and plausible account of environmental holism."[6] None of these studies, however, focus on the possibility of developing an environmental virtue ethics from the Daoist perspective. For some environmental ethicists, environmental virtues have an important and non-eliminative role to play in the development of a fully satisfying environmental ethic. Thus, it is worthwhile to explore the possibility of developing an environmental virtue ethics from the Daoist perspective. This chapter attempts to develop an environmental virtue ethics by arguing for the following two theses:

(P1) Daoism supports an environmental virtue ethic.

(P2) The ethic in question supports a strong version of sustainability.

There are two basic questions that need to be addressed if my argument is to be successful: (1) What is an environmental virtue ethic about? Obviously, we need to make clear what an environmental virtue ethic involves and what aims it needs to achieve if (P1) is to be defensible. (2) How is the concept of sustainability to be understood? We need to address this question simply because for many environmental ethicists, the point of advocating an environmental ethic is to provide "a theoretical platform for censure of environmentally unsustainable practices, policies, and life styles and promotion of environmentally sustainable practices, policies, and life styles."[7] That being the case, to argue for an environmental virtue ethic, we need to make clear what "sustainability" means so as to distinguish environmentally sustainable practices, policies, and life styles from environmentally unsustainable practices, policies, and life styles.

### 1. The Normativity Condition

What is an environmental virtue ethic about? Roughly speaking, an environmental virtue ethic must provide an account of environmental virtue which gives the virtue in question some significant normative force such that the virtue provides ethical standards for censure of environmentally unsustainable practices, policies, and life styles and promotion of environmentally sustainable practices, policies, and life styles. On this conception of an environmental virtue ethic, an ethic which gives no or quite slight normative force to environmental virtue is not an environmental virtue ethic. For instance, an ethic which makes environmental virtue as something optional or contingent upon an individual's personal choices or desires is not an environmental virtue ethic at all. Nor is an ethic which makes environmental virtue contingent upon the stages of economic development of society. Thus, a genuine environmental virtue ethic must give sufficient weight to the normativity of environmental virtue. Let us call this "the normativity condition," a condition which any genuine environmental virtue ethic has to meet. A defense of the condition is that using "environmental virtue ethic" to label an environmental ethic that gives no normative force to environmental virtue is as misleading as to label an ethic that gives no normative force to virtue as "a virtue ethic."

There are two approaches to explaining the normativity of environmental virtue, namely, the external goods approach and the substantive approach.[8] On the external goods approach, any genuinely virtuous agent will exhibit environmental sensitivity and be disposed to promote ecosystem sustainability. The external goods approach argues this on the grounds that ecosystem sustainability is a necessary external good for the cultivation of virtue or the accomplishment of human flourishing. The line of reasoning of the external goods approach can be roughly characterized as follows: Ecosystem sustainability is a precondition of the cultivation of virtue or human flourishing. It is because to cultivate virtue in a person or to accomplish human flourishing presupposes physical conditions such as oxygen, water, food, and fiber that sustain the person as a living being over time. Also, to cultivate virtue in a person or to accomplish human flourishing requires us to ensure the necessary conditions for the cultivation of virtue or the accomplishment of human flourishing. Thus, to cultivate virtue or accomplish flourishing requires us to ensure ecosystem sustainability.[9]

The other approach is the substantive approach. The substantive approach attempts to justify the normativity of environmental virtue by incorporating environmental concern and practice into the substance of the virtues, instead of as a precondition for the cultivation of virtue or accomplishment of human flourishing. On this approach, the cultivation of environmental virtue is a consequence of virtue and not an external means to the development of human virtue or human flourishing. In this regard, the substantive approach has to provide an appropriate account of the virtues such that environmental virtue is considered not as an external means but as an essential aspect of human virtue. Consequently, on this approach, the account of the virtues must be filled out in such a way that a virtuous agent must also be an environmentally virtuous agent. The account will not allow the agent to act anti-environmentally or with environment indifference. Nor will it permit consumptive or degradative human preferences to overly influence the specification of the virtues.[10] One important way to fill out such an account of the virtues is to begin with a thick conception of virtue which, in specifying the virtues, appeals to the metaphysics, epistemology or psychology of a moral tradition. For instance, an environmental virtue ethicist working from the Confucian tradition may explain the normativity of environmental virtue by appealing to the thick conception of virtue of Confucianism. On the Confucian conception of virtue, *jen* or humanness is the first and of utmost importance virtue. This supreme Confucian virtue requires individuals not only to extend sympathy to their human fellows but also to extend moral consideration to other things such as animals and plants. One, therefore, may justify the normativity of environmental virtue in terms of the supreme Confucian virtue.

I do not propose to make a decision on which of these two approaches is better. I just want to show how the normativity of environmental virtue can be explained. As I have shown in the above, a genuine environmental virtue ethic must give sufficient weight to the normativity of environmental virtue. It is useful to discuss how such a normative weight can be given to environmental virtue. In section three, I will argue that Daoism does possess a thick conception of virtue which provides an explanation of the normativity of environmental virtue.

## 2. The Concept of Sustainability

Let us now turn to the question concerning the concept of sustainability. Some environmental ethicists insist that an adequate environmental ethic must meet the condition that it ensures consistent and justified critique of environmentally unsustainable practices and polices. For instance, Ronald Sandler has made the following observation:

> Most environmental ethicists would accept some likeness of the following adequacy condition:
>
> *General Adequacy Condition*: If an environmental ethic is an adequate environmental ethic then it provides a theoretical platform for censure of environmentally unsustainable practices, policies, and life styles and promotion of environmentally sustainable practices, policies, and life styles.[11]

I have no qualm about the idea that a plausible environmental ethic should meet the condition of adequacy as specified above, depending on how the condition is understood. Taking it literally, the condition presupposes that we have consensus on what "environmentally sustainable practices, policies, and life styles" means. The notion of sustainability, however, is notoriously ambiguous and highly contentious. In what follows, I will use three cases to show that there is no consensus on what "sustainability" means among environmental ethicists. On my view, the concept of sustainability like concepts such as "justice" and "equality" is contentious by nature. That being the case, what practices, policies, and life styles should be counted as "environmentally sustainable" depends on what ethical framework we have. If I am right about this, then the above condition of adequacy is boiled down to two requirements: (1) The environmental ethic in question should provide criteria to distinguish environmentally sustainable practices, policies, and life styles from environmentally unsustainable practices, policies, and life styles. (2) It should be able to provide justification for those criteria.

**Case #1.** One influential definition of the notion of sustainability was put forward in the 1987 report of the World Commission on Environment and Development (WCED) entitled "Our common future," which is commonly referred as the Brundtland report. The WCED report defines sustainable development as "development that meets the needs of the present without compromising the ability of future generations to meet their own needs."[12] In other words, for WCED, "sustainability" means sustainable development which is conceived as an important goal that places constraints on the socio-economic and technological processes of a society or the global community such that people of different generations, present or future, are able to meet their own needs to a satisfactory level. Accordingly, it places a moral responsibility on each generation to ensure that future generations inherit a bank of natural resources adequate for satisfying their needs. To achieve this goal, the WCED argues, requires fundamental changes in our practices in agriculture, energy, forestry, and other physical and industrial systems. It requires people to change their attitudes, values, and even lifestyles.

The term sustainability as defined in the WCED report, while having the merit of simplicity, is far from clear and uncontroversial. Some people find it contains inherent difficulties. Vandana Shiva, for instance, argues that the term sustainability as used by the WCED loses its important prescriptive meaning when it is linked with the concept of development, acquiring its full meaning only in the context of the modern market economy. For Shiva, development is just another term for "economic growth and commercialization," which assigns supremacy to the market economy and to its organizing principle based on profits and capital accumulation. It is exactly this concept of development that is at the root of the ecological crisis that we are presently encountering. That being the case, sustainability as understood in the form of sustainable development is deprived of its real meaning. The real meaning of sustainability, according to Shiva, requires us to take seriously "maintaining the integrity of nature's processes, cycles and rhythms."[13]

My comment on the disagreement between WCED and Shiva is as follows. To argue for the "real" meaning or "true" definition of a term, in this case "sustainability," more often than not ends up putting forward some sort of persuasive definition in disguise. To insist on the employment of the real meaning of sustainability, then, is just another way of expressing some set of value judgments on environmental issues. To argue, for instance, that the true definition of sustainability must give primacy to maintaining the integrity of nature's processes, cycles, and rhythms is merely another way of expressing the value judgment affirming such primacy. On the other hand, to argue for the definition that gives primacy to economic growth is no more than another way of expressing a different value judgment.

**Case #2.** Among environmental philosophers, there has been a long existing debate about the two different versions of sustainability, namely, the weak and the strong versions. What is at stake in this debate is whether, or to what extent, substitution between the natural and the physical capital stock should be restricted or allowed. According to the weak version, no limitation should be imposed on substitution between the natural and the physical capital stock, since both forms of capital stock are regarded as complementary. Underlying this weak version of sustainability is the idea that as long as future generations are as well off as current generations, there should be no qualms about the substitution between the two capital stocks. The strong version, on the other hand, maintains that future generations should not only be as well off as current generations, but also should be endowed with minimum, or even the same, level of natural capital stock. That being the case, substitution between the two forms of capital stock should not be unlimited. It is evident that the two versions of sustainability adhere to different moral positions. The weak version presupposes that future generations will not be put in a worse position in terms of the total amount of capital stock. The strong version, however, presupposes that future generations should be endowed with enough natural capital stock.[14]

The dispute over the two versions of sustainability among environmental ethicists reveals the fact that the notion of sustainability used by environmental philosophers is far from neutral. The difference between the two versions of sustainability can be traced to the difference between two moral positions which imply different sets of moral priorities which govern how the natural capital should be used and, therefore, impose different sets of constraints on individuals' activities. In a nutshell, the conceptions of sustainability proposed in environmental disputes are morality laden. Thus, the goal of sustainability is not one that we can take for granted but one that we need to assess and argue for.

**Case #3.** Brian Barry has provided a formal definition for explaining the notion of sustainability:

> The core concept of sustainability is . . . that there is some X whose value should be maintained, as far as it lies within our powers to do so, into the future generation.[15]

Barry's definition of "sustainability" as quoted above is obviously a formal one. It says no more than that sustainability is "the conservation of what matters for future generations."[16] This leaves open for dispute what should be considered as things that matter for future generations. That being the case, the content of sustainability crucially depends on what we think matters for future generations. To put it differently, what "sustainability" means depends on how X should be read. Without further specification of the content of X, we cannot tell what criterion should be used to distinguish between environmentally sustainable practices and policies and the unsustainable ones. In other words, what we need is a substantive definition of the notion of sustainability instead of a formal one if it is to serve our purpose. Barry has considered two candidates for the content of X, namely, utility which is understood as satisfaction of wants or preferences and some notion of equal opportunity. He rejects the wants-satisfaction criterion but accepts some notion of equal opportunity as the appropriate candidate for the content of X. Barry has put forward an argument for rejecting the wants-satisfaction criterion

> The obvious objection to this criterion is that wants are (quite reasonably) dependent on what is, or is expected to be, available. Perhaps people in the future might learn to find satisfaction in totally artificial landscapes, walking on the astroturf amid the plastic trees while the electronic birds sing overhead. But we cannot but believe that something horrible would have happened to human beings if they did not miss real grass, trees, and birds. The want-satisfaction criterion does not enable us to explain what would be wrong with such a world.[17]

Whether one will accept Barry's critique of the wants-satisfaction criterion largely depends on what theory of goodness that person holds. A hedonist, for example, may not find living in totally artifical landscapes as problematic since as hedonists, they accept pleasure being the only intrinsically valuable thing in the world. So if Barry's argument is to be successful to convince the hedonist, it needs some more premisses to explain what is wrong with the Hedonist's view about goodness. Barry does provide some further consideration which by implication would call for the rejection of the hedonist view about goodness as the criteron of what matters for future generations:

> [I]t might appear that what should be maintained for future generations is their chance to live a good life as we conceive it, but even if "we" agreed on what that is (which is manifestly not the case), this would surely be an objectionable criterion for "what matters." For one of the defining characteristics of human beings is their ability to form their own conceptions of the good life. It would be presumptuous – and unfair – of us to pre-empt their choices in the future . . . . We must respect the creativity of people in the future. What this suggests is that the requiremnt is to provide future generations with the [equal] opportunity to live good lives according to their conception of what constitutes a good life.[18]

I am not going to dispute with Barry's argument for using equal opportunity as the criterion of what matters for the future generations. But to note that Barry's argument is even more contentious than the notion of sustainability. For the communitarians, the above argument presupposes a Rawlsian thesis which takes the primacy of the right over the good. That thesis is the one that the communitarians regards as the source of difficulty of Rawlsian liberalism.

I hope that the above three cases suffice to show that there is no consensus about what "sustainability" means among environmental ethicists, and that the concept of sustainability is contentious by nature. That being the case, what practices, policies, and life styles should be counted as "environmentally sustainable" depends on what ethical framework we have. If I am right about this, then the concept of sustainability is not one that we can take for granted but one that we need to assess and argue for. To require an environmental ethic to meet the sustainability condition of adequacy is to require it to provide an adequate conception of sustainability. In order to provide an adequate conception of sustainability, an environmental ethic has to provide a set of criteria to distinguish environmentally sustainable practices, policies, and life styles from environmentally unsustainable practices, policies, and life styles, and provide justification for those criteria.

## 3. A Daoist Environmental Virtue Ethic

In the introductory section, I have said that I will argue for the following two theses:

(P1) Daoism supports an environmental virtue ethic.

(P2) The ethic in question supports a strong version of sustainability.

Also, in section one, I have shown that a genuine environmental virtue ethic must give sufficient weight to the normativity of environmental virtue. And in section two, I have shown that to meet the condition of adequacy an environmental ethic has to be able to do two things: (1) to provide criteria to distinguish environmentally sustainable practices, policies, and life styles from environmentally unsustainable practices, policies, and life styles; (2) to provide justification for those criteria. In what follows, I will show the truth of (P1) by arguing that Daoism holds a thick conception of virtue and that the cultivation of environmental virtue is a logical consequence of this thick conception of virtue. This in turn will show that the Daoist conception of virtue does provide criteria to distinguish environmentally sustainable practices, policies, and life styles from environmentally unsustainable practices, policies, and life styles. Also, I will show the truth of (P2) by arguing that the sustainability criteria provided by the Daoist conception of virtue favors a strong version of sustainability.

Let us focus on the Daoist conception of virtue first. Daoism advocates a number of virtues among which naturalness is regarded as the first and of utmost importance virtue. Naturalness as a first virtue for Daoism is just like *jen* or humanness as a first virtue for Confucianism. It regulates all other Daoist virtues such as *wuwei* (non-action), tranquility and emptiness. The following verses, one of the most often quoted passages in *Daodejing*, provides some textual evidence for regarding naturalness as the first virtue in Daoism

> Man models himself on the Earth;
> The Earth models itself on Heaven;
> Heaven models itself on the Tao;
> And the Tao models itself on naturalness. (*Daodejing*, 25)[19]

The above passage requires some explanation. First, the last line says that the *Dao* models itself on naturalness. However, this should not be read as implying that there exists an abstract entity called "naturalness" and the *Dao* models itself on this entity.[20] Instead, it means only that the *Dao* itself exemplifies naturalness, the property that is inherent in all things, including humans, nature (that is the Earth and Heaven), and even the *Dao*. Second, the above passage seems to imply a hierarchical order in which the human-nature and the nature-Dao relations are defined. At the bottom level is the human-nature relation, which is the relation that man models on nature. At the higher level is the nature-Dao relation, which is the relation that nature models on the *Dao*. The nature-Dao relation is easier to understand. It is simply the relation that *Dao* is the ultimate source and the fundamental governing principle of all things including nature. To put it differently, nature manifests the *Dao*. As for the human-nature relation, humans play a more humble role. Nature for human beings is not something that we can dominate or manipulate. Instead, nature is something that we have to learn from, that is, something like our teacher instead of our servant. What we can learn from nature, however, is not merely scientific facts and natural laws but also the naturalness of the *Dao*, a virtue which we need to develop in us. This is so because given the fact that *Dao* is inherently beyond the bounds of human conception, humans can find the traces of the *Dao* only through this natural world. The human-nature relation, then, is much more like a student-teacher relation. As the student-teacher relation is marked by respect, the Doaist conception of human-nature relation also requires us to show our respect to nature and to cultivate this disposition in ourselves.

At first glance, the environmental virtue derived from the human-nature relation seems to be only an external means to the cultivation of virtue. As one might argue, our respect to nature is only a precondition of the cultivation of virtue if the environmental virtue in question is not a logical consequence of the Daoist virtues; and a virtuous person, therefore, is not necessarily an environmentally virtuous person. The point of the argument is that the environmental virtue derived from the human-nature relation is at most a weak one if the cultivation of virtue does not guarantee the cultivation of environmental virtue. In what follows, I will make two points so as to reply the above query. First, from the Daoist perspective, the human-nature relation that nature to humans is a great teacher is not a contingent relation but the one which reflects a permanent human condition. This is because it is a permanent human condition that our knowledge of the virtue of naturalness and how to develop the virtue in humans largely depends on the teaching of nature. Second, the cultivation of environmental virtue is a logical consequence of the Daoist conception of virtue. This is because, according to the Daoist perspective, to be virtuous requires one to support the naturalness of all other things. Let us look at a frequently cited passage in *Daodejing*

> He who takes an action fails.
> He who grasps things loses them.
> For this reason the sage takes no action and therefore does not fail.
> He grasps nothing and therefore he does not lose anything . . .
> The sage desires to have no desire.
> He does not value rare treasures,
> He learns to be unlearned, and returns to what the multitude has missed.
> Thus he supports the naturalness of ten thousand things and dares not take any action. (*Daodejing*, 64)[21]

For Daoism, the sage is the one who exemplifies the Daoist virtues. It is then clear from the above passage that a virtuous person must support the naturalness of all other things. This implies that a virtuous person must also be an environmentally virtuous person. To see this, we only need to make explicit the content of "naturalness." On one influential interpretation, "naturalness" refers to a state which "indicates the condition when a thing is what it is by itself without any external impulse or interruption."[22] This implies that supporting the naturalness of all things requires taking a non-interruptive approach to all things including nature. In *Daodejing* this non-interruptive approach to all things is coined as "*wuwei*" (or non-action). It then follows from this interpretation of "naturalness" that the virtue of *wuwei* disallows exploitive policies or practices which lead to the degradation of natural environment. For instance, it disallows overly consumptive practice which is the chief contributor to the degradation of natural environment. Thus, it is clear from the above that from the Daoist perspective a virtuous person must also be an environmentally virtuous person. That being the case, environmental virtue is a logical consequence of the virtue of *wuwei* (or non-action).

## 4. Conclusions

In the previous section, I have shown that the human-nature relation in Daoism implies a relation which is analogous to the student-teacher relation and that since the student-teacher relation is marked by respect, the Doaist conception of human-nature relation requires us to respect nature and to cultivate this disposition in ourselves. I also have shown that the virtue of *wuwei* (or non-action) requires adopting a non-interruptive approach to nature. From this it follows that the cultivation of environmental virtue is logically required by the virtue of *wuwei* (or non-action).

I will conclude this chapter by discussing the implication of the above conclusions for sustainability. In what follows, I will argue that the Daoist environmental virtue ethic favors a strong version instead of a weak version of sustainability. I have briefly contrasted these two versions of sustainability in section two. To recall, the weak version imposes no limitation on substitution between the natural and the physical capital stock, since both forms of capital stock are regarded as complementary. The strong version, on the other hand, maintains that substitution between the two forms of capital stock should not be unlimited. Obviously, the Daoist environmental virtue would not allow unlimited substitution between the natural and the physical capital stock. It is because unlimited substitution between the natural and the physical capital stock would result in extensive interruption of nature and that would be disallowed by the virtue of *wuwei* (or non-action), a Daoist virtue which requires adopting a non-interruptive approach to nature.

## NOTES

1. Paul R. Goldin, "Why Daoism is not Environmentalism," *Journal of Chinese Philosophy*, 32:1 (March 2005), p. 76. For the first line of reasoning, see also Ramachandra Guha, "Radical American Environmentalism and Wilderness Preservation: The Third World Critique," *The Ethics of the Environment*, ed. Andrew Brennan (Brookfield, Vt.: Dartmouth, 1995), pp. 239–252.

2. *Enviornmental Ethics*, 8:3 (Winter 1986); *Philosophy East and West*, 37:2 (April 1987); J. Baird Callicott and Roger T. Ames, eds., *Nature in Asian Traditions of Thought: Essays in Environmental Philosophy* (Albany, NY: State University of New York Press, 1989); Norman. J. Girardot, James Miller and Liu Xiaogan, eds., *Daoism and Ecology: Ways within a Cosmic Landscape* (Cambridge, Mass.: Center for the Study of World Religions, Harvard University Press, 2001).

3. Chun-ying Cheng , "On the Environmental Ethics of the *Tao* and the *Ch'I*," *Environmental Ethics*, 8 (1986), pp. 351–370.

4. Roger T. Ames, "Taoism and the Nature of Nature," *Environmental Ethics*, 8 (1986), pp. 317–350; and Kirill Thompson, "Taoist Cultural Reality: The Harmony of Aesthetic Order," *Journal of Chinese Philosophy*, 17 (1990), pp. 175–186.

5. Sharon Rowe and James D.Sellmann, "An Uncommon Alliance: Ecofeminism and Classical Daoist Philosophy," *Environmental Ethics*, 25:2 (Summer 2003), pp. 129–148.

6. Karyn L. Lai, "Conceptual Foundations for Environmental Ethics: A Daoist Perspective," *Environmental Ethics*, 25:3 (Fall 2003), pp. 247–266.

7. Ronald Sandler, "The External Goods Approach to Environmental Virtue Ethics," *Environmental Ethics*, 25:3 (Fall 2003), p. 280.

8. *Ibid.*, pp. 279–293.

9. *Ibid.*, pp. 282–291.

10. *Ibid.*, pp. pp. 291–293.

11. *Ibid.*, p. 280.

12. World Commission on Environmental and Development, *Our Common Future* (Oxford: Oxford University Press, 1987).

13. Vandana Shiva, "Recovering the Real Meaning of Sustainability," *The Environment in Question: Ethics and Global Issues*, eds. David E. Cooper and Joy Plamer (London: Routledge, 1992), pp. 187–193.

14. Robert Ayres, *et al.*, "Strong versus Weak Sustainability: Economics, Natural Sciences, and Consilience," *Environmental Ethics*, 23:2 (Summer 2001), pp. 155–168.

15. Brian Barry, "Sustainability and Intergenerational Justice," *Environmental Ethics: An Anthology*, eds. Andrew Light and Holmes Rolston III (Oxford: Blackwell, 2003), p. 491.

16. *Ibid.*, p. 493.

17. *Ibid.*, p. 491.

18. *Ibid.*, p. 492.

19. Dim Cheuk Lau, trans., *Tao Te Ching*, revised edition (Hong Kong: The Chinese University Press, 1989).

20. Xiaogan Liu, "Naturalness (*Tzu-jan*), the Core Value of Taoism: Its Ancient Meaning and Its Significance Today," *Lao-tzu and the Tao-te-ching*, eds. Livia Kohn and Michael LaFargue (Albany, NY: State University of New York Press, 1998), pp. 211□228.

21. Dim Cheuk Lau, trans., *Tao Te Ching*.

22. Xiaogan Liu, "Naturalness (*Tzu-jan*), the Core Value of Taoism," p. 221.

# Nine

# HEALING AND THE EARTH: DAOIST CULTIVATION IN COMPARATIVE PERSPECTIVE

Livia Kohn

In the Daoist world, human beings are an integral part of nature and the greater universe, which functions in perfect harmony and is fundamentally good. Created in a series of transformations without a radical break from the pure, formless *Dao* (道), the universe manifests itself in successive constellations of an energetic life force known as *qi* (氣) that ideally works to constitute a cosmos of perfect goodness. *Qi* is a bioenergetic potency that causes things to live, grow, develop, and decline. People as much as the planet are originally equipped with primordial *qi* that connects them to the greater universe, but they also work with postnatal or interactive *qi* which can enhance or diminish their primordial potency. What people do socially and in their personal lives, that bears a direct impact on their environment and on the fate of the earth – just as the environment influences how they live and work.

When nature and *qi* work together smoothly, human beings intuit the fundamental cosmic goodness as a sense of well-being and inner harmony, and they often express it in moral rules and environmental ethics. Both personal morality and environmental considerations are part of the cosmic harmony which Daoists embody, and their being in the world increases the overall quality of life. Daoists hope to help people realize that they ultimately are essential to the functioning of the cosmos, and have the power to make it move in harmony or go out of balance. They model ways of acting more and more in cosmic goodness, activating a moral attitude that abstains from harmful actions and an environmental ethics that supports the community of nature and all life. Through their practice and teaching, they inspire people to replace negative and destructive attitudes with supportive and helpful ones, develop a greater awareness of cosmic patterns, and come to create good fortune for themselves, the society around them, and the natural world.

In their cultivation practice, Daoists activate this fundamental understanding in different ways. For one, Daoists are conscious of their physical environment. They place their practice in the ideal constellation of natural settings and make a conscious effort to connect to the energies of the universe. For another, they are aware of energetic changes in the course of day and night, summer and winter, and take care to practice at the right times and make proper use of the powers when available. On a third level, they protect the natural environment and behave socially in an ethical and compassionate way, for avoiding any energetic disturbances.

The Daoist awareness of energetic patterns in time, space, and nature can be compared to the modern scientific understanding of biomagnetic fields and bioelectricity as forming a living matrix that pervades the human body as much as the earth. The Daoist vision of the utterly integrated and perfectly functioning universe, which is similar to the way modern science describes the working of the human body on the cellular level, while our society with its continued assault on both bodies and nature appears like a cancerous growth that can only lead to destruction.

## 1. *Dao* and *Qi*

The most fundamental concept of Daoist environmental ethics is *Dao* or "Way." It can be understood metaphysically as the underlying source and power of the universe, practically as the way in which the world functions, or analytically as the way in which people can (or cannot) speak about reality. As expressed in the ancient classic *Daodejing* (《道德經》*The Book of the Dao and Its Virtue*), *Dao* cannot be named or known with ordinary human senses. It may be described as lying at the root of creation and the cycles of nature, the "mother" of all that keeps nature and society in harmony. *Dao* is a mystical power of universal oneness and a fundamental ontological entity or absolute truth. The intellectual historian Benjamin Schwartz describes it as "organic order" – "organic" in the sense that it is part of the world and not a transcendent other as in Western religion, "order" because it can be felt in the rhythms of the world, in the manifestation of organized patterns (Schwartz, 1985, p. 215).

Another way to think of *Dao* is as two concentric circles, a smaller one in the center and a larger on the periphery. The dense, smaller circle in the center is *Dao* at the root of creation – tight, concentrated, intense, ultimately unknowable, ineffable, and beyond conscious or sensory access. The looser, larger circle at the periphery is *Dao* as it appears in the world, the patterned cycle of life and visible nature. Here we can see *Dao* as it comes and goes, rises and sets, rains and shines, lightens and darkens – the ever-changing yet everlasting alteration of natural patterns, yin and yang, life and death. This *Dao* is what people and rulers need to adapt to; they should go along with it to create harmony and be at ease with it to find fulfillment (Kohn, 2001, p. 20; Henricks, 2000, A:7).

The inner, central *Dao* at the root of creation, on the other hand, is ultimately beyond the activities of daily existence. It maintains the outer circle but does not do anything in itself. Finding intuitive access to this inner *Dao* helps people to find harmony with its peripheral manifestations and to attune to universal harmony on a deeper level. Aligning oneself with the root of all empowers the self and aids society in the development of utmost balance.

*Dao* is always good. Bad times, bad things, bad people, all forms of evil happen when things move against the flow of *Dao*. That does not mean that there is no room for recession, decline, or death. All these things are there, but not considered evil. Instead, they are a necessary part in the ongoing flux of life in which everything is relative and related to everything else. Yin and yang are interdependent and relative forces that always move together and in alteration, and neither of them is evil. Evil occurs when violent decline is forced upon nature or society at a time of growth, or when massive expansion is pushed forward at a time of rest or reduction. It is essential to know the patterns of *Dao* – personally or politically – and learn to adapt to its rhythms.

The way to be with *Dao* is through non-action (*wuwei* 無為) and naturalness (*ziran* 自然). This does not mean doing nothing or being totally spontaneous without any planning. It means letting go of egotistic concerns and passions and desires on the personal level, finding a sense of where life, nature, and the world are headed on the social level, and abstaining from forceful and interfering measures in the political realm (see Liu, 1991; 1998). As the *Daodejing* says:

> Act on things and you will ruin them.
> Grasp for things and you will lose them.
> Therefore the sage acts with non-action and has no ruin,
> Lets go of grasping and has no loss.
> (ch. 64; Henricks, *Guodian*, A:6)

The best way to achieve non-action is by adjusting to the concrete aspect of *Dao*, the universal life force of *qi*. *Qi* is the material energy of the universe, the basic stuff of nature. In ancient sources it is associated with mist, fog, and moving clouds. The character for *qi* as it appears in the oracle bones of the Shang Dynasty (1766-1122 BC), consists of two parts: an image of someone eating and grain in a pot. Combined, these parts signal *qi*, the quality which nourishes, warms, transforms, and rises. *Qi*, therefore, is contained in the foods we eat and the air we breathe. But more subtly it is also the life force in the human body and as such is the basis of all physical vitality (Kohn, 2005, p. 11).

By extension, *qi* also denotes anything perceptible but intangible: atmosphere, smoke, aroma, vapor, a sense of intuition, foreboding, or even ghosts. There is only one *qi*, just as there is only one *Dao*. But it, too, appears on different levels of subtlety and in different modes. At the center, there is primordial *qi*, prenatal *qi*, or true, perfect *qi*; at the periphery, there is postnatal *qi* – like the measurable *Dao* it is in constant motion and divided according to categories such as temperature, density, speed of flow, and impact on human life.

*Qi* is the basic material of all that exists. It animates life and furnishes functional power of events. *Qi* is the root of the human body; its quality and movement determine human health. *Qi* can be discussed in terms of quantity, since having more means stronger metabolic function. This does not mean that health is a byproduct of storing large quantities of *qi*. Instead, there is a normal or healthy amount of *qi* in every person and every aspect of nature, and health manifests in their overall balance and harmony, moderation and smoothness of flow. This flow is envisioned as a complex system of waterways both in nature and in the human body. In the latter, the "Ocean of *Qi*" is in the abdomen; rivers of *qi* flow through the upper torso, arms, and legs; springs of *qi* sprout at the wrists and ankles; and wells of *qi* are found in the fingers and toes. In nature and the body, even a small spot in this complex system can influence the whole, so that overall balance and smoothness are the general goal.

Human and natural life is the accumulation of *qi*; death is its dispersal. After receiving a core potential of primordial *qi* at birth, people throughout life need to sustain it. They do so by drawing postnatal *qi* into the body from air and food, and from other people through sexual, emotional, and social interaction. But they also lose *qi* through breathing bad air, living in polluted conditions, overburdening or diminishing their bodies with food and drink, or getting involved in negative emotions and excessive sexual or social interactions.

Health in the universe is not just the absence of symptoms and ailments. It is the presence of a strong vital energy and of a smooth, harmonious, and active flow of *qi* in a steady alteration of yin and yang according to the system of the five phases which are symbolized by five material objects:

| minor yang | major yang | yin-yang | minor yin | major yin |
|---|---|---|---|---|
| wood | fire | earth | metal | water |

These five continue to produce each other continuously in a harmonious cycle in the order presented. *Qi* that flows in this proper manner and in the right amount is known as the state of "proper *qi*" (*zhengqi* 正氣) or also translated as "upright *qi*." Found in personal health, this is matched by health in nature, defined as regular weather patterns and the absence of disasters. It is also present as health in society in the peaceful coexistence among families, clans, villages, and states. This harmony on all levels, the cosmic presence of a steady and pleasant flow of *qi*, is what the Chinese call the state of Great Peace, a state venerated by Confucians and Daoists alike (see Eichhorn, 1957).

The opposite of *zhengqi* is "wayward *qi*" (*xieqi* 邪氣) or also called "deviant *qi*," "pathogenic *qi*," "heteropathic *qi*," or "evil *qi*." All these expressions are used in Western textbooks to translate the same Chinese term. The variety reflects the different views of the translators and shifts the meaning of the term and thereby the understanding of what goes on in the body. Typically, medical or Latin-based words like "heteropathic" tend to be more technical and overshadow the moral implications of the original term, while words like "evil" or "deviant" have moral and social instead of medical implications. Both are equally present in the original, reconfirming the basic Chinese and Daoist view of environmental ethics and personal health as part of the same essential phenomenon.

"Wayward *qi*" *Xieqi* is *qi* that has lost the harmonious pattern of flow and no longer supports the dynamic forces of change. Whereas *zhengqi* moves in a steady, harmonious rhythm and effects daily renewal, supporting health and long life, *xieqi* is disorderly and dysfunctional, and creates change that violates the normal order (see Kaptchuk, 1983). When it becomes dominant, the *qi*-flow can turn upon itself and deplete the body's resources. As any sick person, decimated forest, or manmade edifice no longer operates as part of a universal system and is not in tune with the basic life force. *Xieqi* appears when *qi* begins to move too fast or too slow, is excessive or depleted, or creates rushes or obstructions. It disturbs the regular flow and causes physical ailments, natural disasters, and social unrest.

*Qi* can become excessive through outside influences such as too much heat or cold or through inside patterns such as too much emotion or stimulation. Excessive *qi* can be moving too fast or too sluggish, as in the case of excessive dampness. Whatever the case, from a universal perspective there is no extra or new *qi* created, but localized disharmonies have arisen because existing *qi* has become excessive and harmful. Still, even describing it in this way we are thinking in terms of *qi* as an energetic substance, which it is not. A better way to speak of it would be to say that the process itself of turning hot or angry is *qi*, that the way things move and change is what constitutes being *qi* (Kohn, 2005, p. 13).

Similarly, *qi* can be in depletion. This may mean that there is a tighter flow of *qi* or that the volume and density of *qi* have decreased, which is the case in serious prolonged illness or complex environmental issues. However, more commonly it means that the *qi* activity level is reduced, that its flow is not quite up to standard, that there is a lower than normal concentration of *qi* in one or the other aspect. In this case, too, the body, society, and the natural world are adversely affected, and the imbalance spreads from one to the other.

The concepts of *Dao* and *qi* in Daoism, which formulate just how intricately related people and the environment are. Both *Dao* and *qi* are equally important on all different levels of existence and have to maintained properly. Practicing Daoists in this context, then, have the task to model ways of achieving perfect balance and ideal balance of *Dao* and *qi* in human life and the natural world.

## 2. Space

Daoists begin to fulfill this task by closely following the classic Chinese practice of placing oneself most auspiciously into the physical setting. Known as *Fengshui* (風水) means "wind and water" and sometimes described as "geomancy," this begins with the recognition of two major forms of *qi*, analogous to the medical diagnosis of excess and deficiency. They are *shaqi* and *siqi*. *Shaqi* (殺氣) means "killing *qi*," is fast moving or rushing *qi*. It moves along straight lines and is found in pointed arrows and angular forms. Moving very fast, it takes away health and good fortune. An example is the placement of a house at the top of a T-junction. The *qi* from the road rushes straight at it and through it, taking positive energies away with it. But even ordinary entrances to a house or compound are in danger and should be protected. As a result, in traditional China the main entrance to a compound is usually blocked off to the street by the so-called spirit wall: a free-standing wall slightly broader than the gateway that is set inside the main wall right after the gate (Lip, 1995).

Other common locations of *shaqi* include staircases that rise straight up from the front door of the house, long corridors, and a window directly across from the door of a room. In all cases, these long, straight, pathways invite the *qi* to rush through, taking good fortune with it. It can be slowed down and mellowed with the help of a screen, plant, or curtain. Then there are the sink and toilet, where water flows out of the house, another drain of *qi*. In traditional China, bath houses and outhouses were completely separate from the dwelling and placed in their *Fengshui* environment. In modern apartments, bathrooms and open toilets contribute to the overly rushing nature of life, as does the tendency to disregard *Fengshui* rules and to place key items, such as the stove, bed, or desk in a *shaqi* location (Rossbach, 1983).

The other major form of *qi* is *siqi* (死氣) means "dead *qi*." Blocked or obstructed *qi*, it is deficient because it cannot move freely. *Siqi* is found in all types of clutter – whether unused items or an overabundance of appreciated possessions, such as too many plants, books, dishes, clothes, and the like. *Fengshui* clutter does not have to be in the house itself; even piles disregarded stuff stored in other locations can have a negative impact, and any assembly of material objects that does not let *qi* pass through freely is detrimental to one's well-being and should be cleaned out. Daoists accordingly maintain a stark, clutter-free environment, living an example of purity and open *qi*. They also keep their housing and temples meticulously clean, since another form of blocked *qi* appears in dirty windows, dark corners of the house, and different heavy areas and objects. Often lamps, fountains, mirrors, paintings, and plants are used to punctuate space and provide a soft undulating motion of *qi*, creating a space that is both friendly and peaceful.

The ideal form of *qi*-flow in and around one's residence is one of balance. There should be nice rounded shapes, spacious corners, a pleasant atmosphere, and ideally the presence of all colors of the five phases (green, red, yellow, white, and black) and the different aspects of yin and yang (heavy and light, dark and bright, soft and hard). When all aspects are represented in every room, *qi* can flow harmoniously, and one will find health and good fortune (Rossbach and Lin, 1994).

Daoists use these general guidelines to manage their temples, altars, and houses and place them in a *qi*-rich environment. Ideally any building should be surrounded by rich vegetation and some wildlife, showing that the ground is fertile and the *qi* is prosperous; barren hills and deserts should be avoided. If working in a city-setting, Daoists prefer their sites to be surrounded by greenery and located near parks, playgrounds, health areas, and schools, while avoiding hospitals, cemeteries, power stations, prisons, butcheries, and industrial estates. Most commonly, Daoists tend to avoid congested areas and prefer to locate themselves in the mountains, where they can have their buildings protected by nature, work with a landscape that reflects all the five phases – pointy (fire), tall (wood), angular (metal), undulating (water), and round (earth) – and is active in terms of animal forms, showing resemblance to dragons, tigers, or rabbits (see Field, 2001).

Mountains play a key role in the Daoist management of space (see Hahn, 1988; Taoist Association, 2002). Apart yet essential, they are a feature of earth that yet connects to the heavens above. Mountains are sacred places that contain wondrous plants, animals, and beings who can help the practitioner's quest. They contain grottoes, caverns, caves, underground spaces, mountain hollows and inner corridors that all possess qualities manmade places do not have (Miura, 1983). They are archetypal chambers of reflection and the classical home of the immortals (see Kaltenmark, 1953), where miraculous, inexplicable phenomena abound – such as lush growth despite droughts, or discovery of an ancient scroll within an uninhabited cave.

Mountains, are a natural sanctuary, both as a holy place dedicated to the *Dao* and as a place of refuge and protection (Schipper, 2001, p. 82). Providing a wondrous opportunity to venture into the unknown, they may appear solid, yet are passages of *qi*, allowing smooth air currents and water flows, which reflect the inner workings of the human body. Blood equals water; air equals breath. Spermatic liquids form pools; walls constitute shapes like inner organs or viscera (Hahn, 2000, p. 495). In their self-contained interior, free from the defilements and impacts of the outside world, Daoist adepts can perform alchemical and meditative experiments of self-transformation, thereby to experience a renewed closeness to the *Dao*.

Mountains in many ways are thus the outer body of the self. Daoist hermits, aware of the characteristics and the possibilities of a given locality, resolutely place themselves in charge of all the forces a mountain harbors: water and forests, wild birds and beasts, ghosts and divinities, minerals and ores (Hahn, 2000, p. 694; see also Porter, 1993; Campany, 2001). These forces have to be treated with great respect, and mountains can only be entered with great care, using the proper spells and talismans, and after having undergone thorough preparation and purification (Hahn, 2001, p. 205). As the alchemist and Daoist master Ge Hong (283-343 AD) notes in his *Baopuzi* (《抱朴子》*Book of the Master Who Embraces Simplicity*, Daozang (道藏) or Daoist canon (DZ) 1285):

> All mountains, whether large or small, contain gods and powers, and the strength of these divinities is directly proportional to the size of the mountain. To enter the mountains without the proper spells and talismans is to be certain of anxiety or harm. In some cases people fall ill, are wounded, or become stricken with fear. In other cases, lights and shadows are seen, strange sounds are heard. Lack of the proper talisman can make a large tree fall when there is no wind; or a high rock tumble for no apparent reason, striking and killing people. It can confuse travelers or drive them madly on, so that they fall into ravines. Lack of preparation may cause you to meet with tigers, wolves, or poisonous insects that will harm you. Mountains are not to be entered lightly. (ch. 17; Ware, 1966, pp. 279-280)

Undergoing careful training and venerating the mountain properly, its sacred space becomes the sacred space of the self, allowing and facilitating the ultimate transformation back to the roots of the *Dao*. Earth and body become one, and by living this oneness, Daoists demonstrate the necessity to work with and in nature – a nature that is ultimate not separate from the individual person.

In a yet different mode of activating spatial energies, Daoists ingest the "five sprouts" (*wuya* 五芽), the pure energies of the five directions. Part of medieval Daoist cultivation and described especially in the texts of Highest Clarity (*Shangqing* 上清), the practice begins with swallowing the saliva while chanting invocations to the original *qi* of the four cardinal directions. Then adepts face the direction in question, usually beginning with the east, and in their minds visualize the *qi* of that direction in its appropriate color, green in the case of the east. A general mist in the beginning, the *qi* gradually forms into a ball, like the rising sun, then through further concentration shrinks in size and is made to come close to the adept. Eventually the size of a pill, the sprout can be swallowed and guided mentally to the organ of its correspondence, the liver in the case of the east. A suitable incantation places it firmly in its new receptacle, and gradually the adepts body becomes infused with cosmic energy and partakes more actively of the cosmos as a whole (Robinet, 1989, pp. 165-166).

The sprouts, as Isabelle Robinet points out, are originally the germinal essences of the clouds. They represent the yin principle of Heaven – that is, the yin within the yang. They manifest in human saliva, again a yin element in the upper, yang, part of the body. They help to nourish and strengthen the five inner organs. They are quite tender, comparable to the fresh sprouts of plants; they assemble at dawn in the celestial capital, from where they spread all over the universe until the sun begins to shine. Turning like the wheels of a carriage, they ascend to the gates of the nine heavens, from where they continue to the medium level of the world – to the five sacred mountains ruled over by the five emperors of the five directions – and finally descend into the individual adept. They then pass through the three major levels of the cosmos (Robinet, 1989, p. 166).

The virtue of these sprouts is twofold. Emanations of the highest pole, they are full of the power of far-off regions, the fringes of civilization where the *Dao* resides in a rawer state. At the same time, they contain the entire potential of being in its nascent state. This growth potential, the small and imperceptible *qi* in a state of pure becoming, is the main objective for the Daoist practitioner. "Sprouting" means inherent creation, purity, newness, return to youth. It also implies the prevalence of the soft over the hard and the power of yin over yang that Laozi describes in the *Daodejing*. Here yin is represented by the saliva that adepts absorb. The practice is undertaken at dawn, the time when everything awakens to life, yet another symbol of creative, unstructured potential. By ingesting the sprouts, Daoists partake of the inherent power of celestial bodies and feed on the pure creative energy of the universe, and recovering a sense of primordial oneness and fundamental harmony (see Robinet, 1993).

In contemporary practice, this technique survives with some modifications in an exercise call Primordial Qigong (*Wuji Qigong* 無極氣功), which was transmitted to the West from the Daoists of Mount Wudang in Hunan by Healing Dao master Michael Winn. Named after the Non-Ultimate at the beginning of creation, in involves the active intake of heavenly and earthly *qi*. Practitioners begin by facing east and breathing deeply, then raise and lower their arms and rotate their hands as they encourage the *qi* to join their bodies. They complete twelve movements of Earth, repeat these going counterclockwise in all four directions, then perform ten movements of Heaven, also in all directions. Once one cycle is complete, ending in the south, it begins again, rotating the practice clockwise around the four directions back to the east. The practice is deeply meditative. As Michael Winn says in his beginning instructions:

> Stand facing east, the direction of sunrise and new beginnings. Open your heart to change. Inwardly smile or call to the beings of all directions, focusing on the four directions and three realms of Heaven above, Earth below, and the Human in the center. Hold the intention to return to your core being, to maintain centeredness in life, and balance all perceptions arising in the present moment.

> Open yourself to being in *wuji*, the "supreme unknown," the primordial space. It represents the origin or primal matter at the center of all sacred space. Ask to create your life from this center, to accept its power of unconditional acceptance, and to solve any problems. . .
>
> To conclude, bring both arms back and around in a wide circle. Rest and breathe deeply three times, visualizing the *qi* of Heaven and Earth flowing through you as a bright golden light. Feel warm and protected, at home in the cosmos. (*www.taichi-enlightenment.com*)

In all these different ways, by following the rules of *Fengshui*, by placing themselves into the cosmically empowered space of the sacred mountain, by ingesting the five sprouts, and by absorbing the *qi* of the four directions, Daoists thus activate a conscious relationship to the environment as it manifests in space. They work with space in its manmade, natural, and bodily manifestations, integrating the three forms and in all cases allowing the *qi* of nature and self to be pervasive, open, fluid, and free. They create beauty and harmony in their practice, realizing in concrete activities the oneness of body, nature, and the *Dao*.

### 3. Time

Just as Daoists place themselves ideally in space, so they also work closely with time, again beginning with their adaptation of general Chinese practices and following the traditional Chinese calendar. Like ours, the Chinese distinguish four seasons – with sometimes, to match the five phases, a fifth, a kind of Indian summer, added between summer and fall – marked by the solstices and the equinoxes. Unlike in our calendar, these markers are not considered the seasons' beginnings. Instead, they are the high points of the seasons, which begin about six weeks prior to them. This system of having eight major cadences in the year – two solstices, two equinoxes, and four seasons' beginnings – is known as the Eight Nodes (*bajie* 八節), and roughly matches the festivals of Western pagan or Wiccan religion: Candlemas (New Year, 2 February), Ostera (Spring Equinox, 21 March), Beltane (1 May), Summer Solstice (21 June), Lammas (2 August), Fall Equinox (21 September), Samhain (Halloween, 31 October), and Winter Solstice (21 December). In addition, the Chinese divide their year into twenty-four solar periods of about two weeks each, which are named after weather patterns such as "great heat," "slight cold," "great rain," and "slight snow," but also include the solstices and equinoxes (Kohn, 2001, p. 47).

Adaptation to the seasonal cycle has been a key concern in Chinese culture since antiquity. First formulated in the "Monthly Commandments" of the *Liji* (《禮記· 月令》 *Book of Rites*, "Yueling," see Legge, 1966) and later expanded in the Han Dynasty Daoist collection *Huainanzi* (《淮南子》*Book of the Prince of Huainan*, DZ 1184; see Major, 1993), seasonal rules have played an important part in traditional Chinese culture and politics. For example, it became a general rule that because Heaven and Earth make the myriad creatures blossom in spring, one should in that season sleep and rise early, loosen the hair, relax the body, and allow all beings to live, abstaining from killing. Spring was a time of giving and not of taking, of reward and not of punishment. All executions had to wait until the fall, and slaughter and hunting were severely limited. By following these injunctions, humanity was believed to act in proper alignment with the *qi* of spring and to secure health and harmony for both body and society. Any actions against the dominant *qi*-flow, on the other hand, might cause harm to the liver, arouse aggression and anger, or lead to social upheaval and natural disasters.

While politicians and companies have long abandoned any efforts of seasonal adjustments, Daoists and modern longevity followers are still working closely with the seasons and strive to use the *qi* most appropriately. For example, as described in the medieval *Zhaijie lu* (《齋戒錄》*Record of Purgations and Precepts*, DZ 464; see Malek, 1985), Daoists regularly performed major ceremonies or purgations (*zhai* 齋) at the Eight Nodes, linking the seasonal transitions with the establishment of cosmic goodness and clarity.

In a similar vein, Liu Zhengcai, a modern student of longevity, outlines seasonally appropriate ways of dressing, moving, and eating right in the beginning of his book (1990, pp. 5-10). He emphasizes the alternation of yin and yang and strongly advises people to avoid cooling too much in summer and remaining too warm in winter. Breathing and meditation are recommended to balance the natural patterns of *qi*. "Just rest calmly without any perplexing thoughts and one will arrive at refreshing mountains" (1990, p. 7).

Beyond the seasonal pattern, the Chinese traditional calendar also had a cycle of sixty that was applied to years, months, days, and hours. Formulated in pairs of characters that represented the so-called twelve heavenly stems and ten earthly branches, it goes back to Shang Dynasty astronomy and time calculation. The cycle was formally established in the Han Dynasty and the first "1-rat" (*jiazi* 甲子) year known is the year 4 CE, so that from then onward all dates can be calculated. Each new cycle was seen as a new beginning, based on the idea that human life lasted for approximately sixty years and that, once it was over, a complete renewal occurred. It is also at the root of Chinese fate calculation, where the so-called Eight Characters (two each for year, month, day, and hour of birth) determine a suitable marriage match. Religious rituals, of all Chinese traditions, are scheduled according to the auspicious or inauspicious nature of the signs, and many of the most important Daoist rites occur once every sixty years to mark the renewal of the cosmos. In addition, Daoist and other millenarian movements have focused on certain years, such as *jiazi*, the first year of the cycle, as the starting point of the new world (Kohn, 2001, p. 48).

The sixty-day cycle forms a key aspect of Daoist cultivation. As the Highest Clarity work *Shenxian shiqi jingui miaolu* (《神仙食氣金櫃妙錄》 *Wondrous Record of the Golden Casket on the Spirit Immortals' Practice of Eating Qi*, DZ 836) notes, the cultivation of immortality begins with refining the *qi* through the elimination of gross substances such as solid food. To replace food intake with pure *qi* and thereby make the body subtler and closer to the *Dao*, adepts should ingest a divine talisman every day, changing its text and accompanying spell according to the sixty-day cycle. In each case, a divine master of time in the form of a jade maiden is invoked, using her personal name and making sure of the correct timing (see Inoue, 1992). The text says:

> Begin the practice on a *jiazi* day and continue it for sixty days. Always use the same basic spell, but change the name of the day and the name of the responsible jade maiden as appropriate. Always include her full name when you call on her, but do not change anything else in the spell. The Dao has two patterns, one yin, one yang. Follow the instructions accordingly. Thus, the *jiazi* in this talisman needs to be changed to *bingyin* 丙寅 when the *bingyin* day arrives, and so on for the thirty yang days of the cycle. (1b-2a)

The spell runs:

> I, so-and so, love and relish the Dao of perfection,
> Absorb and eat the *qi* of central harmony.
> Oh, Jade Maiden of Great Mystery of the *jiazi* day,
> Chengyi, come and serve me.
> Wherever I, so-and-so, practice this diet method,
> May the method work perfectly as chosen.
> May it prevent me from feeling hungry and thirsty,
> And let even armies big or small, people few or many
> Be satisfied completely. (1b)

Working with the symbolic and divine powers associated with the sixty-day cycle, Daoists refined the seasonal commandments to a new and more intricate level of time management, adjusting their activities to the cosmos, and using the universal energies to create a more subtle way of being in the world.

The same general pattern also applies to their use of daily rhythms. Practically every Daoist text, from antiquity to today, states firmly that "from midnight to noon is the time of life-*qi*; from noon to midnight is the time of death-*qi*. Always practice during the time of life-*qi*" – it is from the seventh-century *Yangxing yanming lu* (《養性延命錄》*Record of Nourishing Inner Nature and Extending Destiny*), ascribed to the famous Daoist and physician Sun Simiao. Adepts accordingly are admonished to practice after midnight on in the early morning – and this is indeed still common among Taiji and Qigong followers today. An example of a medieval Daoist cultivation practice is found in the *Huanzhen xiansheng fu neiqi juefa* (《幻真先生服內氣訣法》*Master Huanzhen's Formula for Absorbing Internal Qi*, DZ 828), a text of the mid-Tang period or eighth century. It says:

> Everyday after midnight when the time of life-*qi* has come or when you first wake up around the fifth watch, begin by exhaling all the stale and bad *qi* from the belly, using *chui* breath [a sharp expulsion of air]. Do one set of nine repetitions, exhaling from the mouth. . . . At the fifth watch, the day is just breaking and the *qi* is in perfect harmony. Also, your belly is empty which allows optimal practice. ( 1a )

In addition, Daoists use the twelve major periods or double-hours of the day for optimal *qi*-absorption. As specified in the *Shenxian shiqi jingui miaolu* cited earlier, one should practice swallowing the *qi* to replace solid food a certain number of times at specific periods:

| | | |
|---|---|---|
| first light: 7 x 7; | sunrise: 6 x 6; | daybreak: 5 x 5 |
| midmorning: 4 x 4; | noon: 9 x 9; | mid afternoon: 8 x 8 |
| last light: 7 x 7; | sunset: 6 x 6; | dusk: 5 x 5 |
| mid evening: 4 x 4; | Midnight: 9 x 9; | cock crow: 8 x 8 (2a) |

Time in Daoist cultivation is a complex affair that finds expression in seasonal rules, the symbolic and rhythmical use of the sixty-year and sixty-day cycles, and the activation of *qi* according to specific hours of the day. As they do with space, so with time Daoists adapt the common understanding and basic patterns of Chinese traditional culture, then add their own specific ways. Time as much as space is a cosmic mode of *qi*-manifestation and has to be accepted and activated accordingly. It can make all the difference in the effectiveness of specific practices and can lead to immortality and free of disease. A key feature of the cosmos, it has a strong impact on the human body, on life, and on success – and vice versa, human activities and attitudes can distort and confuse the proper rhythm of the seasons, causing the wind and rains to come at the wrong moment, turning day into night and night into day. Keeping with the patterns of nature leads to harmony and oneness, to peace and stability; upsetting them results in difficulties and confusion.

## 4. Ethics

Understanding these basic principles of the cosmic workings, Daoists also have created a strong sense of environmental ethics. Already the earliest communal code, the *Laojun yibai bashi jie* (《老君一百八十戒》*The 180 Precepts of Lord Lao*, in DZ 786), which forms the basis of all communal Daoist codes to the present day (see Kohn, 2004), contains over twenty rules that directly address environmental issues and prohibit the abuse of natural resources. The code lists a variety of regulations regarding many different areas of life and experience that cover all contingencies of community life, from interaction among members through the consumption and treatment of food to personal integrity and spiritual propriety. They prohibit abortion and musical entertainment, the forming of cliques and political parties, ownership of slaves and cruelty to animals. They also pay close attention to personal interaction. For example, members must not spy into the affairs of others, use fancy or ambiguous language, scold others in anger, discuss the faults of others, criticize the teachers, claim to have great merit and virtue, or use words in other unseemly ways (see Penny, 1996; Hendrischke and Penny, 1996; Schipper, 2001).

Beyond this, members are also held to honor the natural environment, to avoid cutting trees, picking flowers, startling birds and beasts, disturbing worms and insects, and burning fields or forests. They should generally live in a low-impact manner, be conservative in their use of food and drink, keep the roads and wells free from obstructions, and abstain from throwing harmful substances into public supplies of water and food. Their building of houses, graves, and roads should match the *qi* of the natural surroundings, and they should respect natural marshes and waterways, only minimally interfering with the topography of the land. Often the rules prohibit the "wanton" taking or despoiling of natural things, and leaving the possibility open to use these resources as and when necessary, with respect and consideration. As Kristofer Schipper points out, the main theme of the rules is "that of *respect*, not only for nature, but for all men, women, and children, their ways of life, customs, and culture" (Schipper, 2001, p. 84).

The environmental rules of text reflect the concerns of an agricultural community of people who need to preserve and manage the land to their best advantage and who live closely together, joined in the mutual enterprise of creating harmony in a difficult world. Located in Sichuan, which at the time was highly advanced and very civilized, these early Daoists had to confront "all the endemic ills connected with sedentary life and high economic pressure, the scourges of rural China: abusive taxation, raids by external nomadic tribes, high population density, epidemics, and famine" (Schipper, 2001, p. 83). It is not surprising that they created a set of rules that protected not only the human community but also the natural world, striving in their communal organization as much as in individual cultivation for an overall harmony in the cosmos and proper alignment with Heaven and Earth.

This alignment with Heaven and Earth, in Daoist thought, is essential for the good fortune of humanity and the optimal realization of the individual's destiny. As outlined in the *Chisongzi zhongjie jing* (《赤松子中戒經》 *Essential Precepts of Master Redpine,* DZ 185) of the fourth century, people consist of the *qi* of Heaven and Earth and their thoughts and actions have a direct impact on the functioning of the world (see Kohn, 2004). The text says:

> Human beings live between Heaven and Earth and are endowed with the two *qi* of yin and yang. Sovereign Heaven, although high, yet has its correspondence down below. Mother Earth, although low, yet has its correspondence far above. Heaven does not speak, yet the four seasons move in order. Earth does not speak, yet the myriad beings come to life. People reside right between the two. All their licentious intentions and passionate desires, whatever they do or do not do, Heaven and Earth know all about it. For this reason we say that Heaven has four-sided [all-round] knowledge. (2a)

This knowledge not only extends to people's overt and open actions but also involves their hidden thoughts and secret intentions. All these have an immediate effect, so that "when the people of the world commit violations, bad actions, or faults, or speak contrary words, Heaven's way is no longer even; instead, it is bent and loses its spontaneity" (5b). All bad deeds, all vicious thoughts, all cursing and scolding upset Heaven and create cosmic imbalance. So does the suffering that people undergo as a result of their bad deeds. Even worse, Heaven is strongly discomfited when people afterwards try to implore it and garner favor. The text says:

> Getting imprisoned and locked into the cangue [punishment board] upsets Heaven. Being hungry and poor, ill and sick upsets Heaven. Unseasonable cold and heat, frost and snow upset Heaven. Irregularities in the length of day and night upset Heaven.
>
> It upsets Heaven even more when people do evil then come to pray and besiege it [for help]. Or go against the four seasons, disobey the five phases, expose their naked body to the Three Luminants [sun, moon, planets], and then come to request a benefit at a shrine or temple of the Three Luminants or the various stars and constellations. (5b-6a)

People cause cosmic disharmony by not having respect for Heaven and Earth, the demons and spirits, by cursing the wind and the rain, by denigrating the sages and the scriptural teachings, and by desecrating shrines and temples

(6a). Not only religious and cosmic violations, but all manner of social harm disturbs the cosmos: being unfilial toward their father and mother; digging up tombs to steal the valuables of the dead; cheating the blind, the deaf, and the dumb; throwing impure substances into food and drink; killing other beings; accusing and slandering others or spying on their affairs; obstructing roads and letting drains be blocked; stealing and cheating, destroying nature's riches, and in other ways harming society and life (6ab). All these acts cause Heaven and Earth to be upset and imbalanced. They in turn do not keep their displeasure a secret but give human beings ample warning that something is amiss. As *The Essential Precepts of Master Redpine* states:

> Heaven never cheats on living beings but shows them its inclinations, like a shadow following a shape. It has day and night, light and darkness, thunder and lightning, rain and snow, intertwining rainbows, eclipses of the sun and the moon, and floating characters of wisdom. All these are signs given by Heaven.
>
> Similarly Earth never cheats on living beings but shows them its inclinations, like an echo following a sound. It makes rivers and streams dry up and brings forth landslides and earthquakes, hurricanes and tornadoes, sandstorms and moving stones, floods and locust plagues, famines and droughts, epidemics and other disasters. All these are signs given by Earth. (2b)

In connection between human behavior and environmental conditions, natural disasters are seen as the direct result of immoral and disturbing attitudes and actions on the side of humanity. Daoists accordingly urge people to understand that they are at the root of Heaven and Earth and have the power to make the cosmos move in harmony or go out of balance. They embody this knowledge themselves, and in their practice strive to act more and more in goodness – a moral goodness that abstains from harmful actions and supports the community of all life. They replace negative and destructive attitudes with supportive and helpful ones, develop an encompassing awareness of Heaven and Earth, and come to create good fortune for themselves, the society around them, and the world at large.

As was the case in the Daoist management of space and time, so in their ethical thinking they exhibit a central concern with *qi* or cosmic energy, which not only keeps individual life together but also determines the social network and the interaction patterns with Heaven and Earth. Essential on all levels – nature, state, and family – *qi* must never be lost but always be guarded, preserved, strengthened, and cultivated. Because each level depends on the other, and there is a basic need of harmony in each; failure of harmony on one level leads to disruptions on another. The 180 community rules of the early groups as much as the abstract ethical thought of the middle ages emphasize the creation and maintenance of harmony and good *qi* on all levels of life, as no one ever stands alone but that all life is always embedded in a multitude of networks and energetic constellations which need to be nurtured to the utmost. Moral action in this context becomes the creation and continuation of the most essential, fundamental web of life, the proper way of relating to everything around one – whether human or nonhuman, animate or inanimate.

## 5. A Scientific Perspective

The Daoist environmental thinking and practices presented were formulated entirely in the concepts and terminology of traditional China, which has to be learned and appreciated for the system to make sense. It would be most beneficial and helpful for inter-religious and intercultural dialogue to be able to translate some of the Daoist ideas into the terms and visions of modern Western science, with the hope that they may have a greater impact on the development of environmental wisdom in the world today. Indeed, there are a few Western concepts, some more established than others, that are amazingly close to Daoist ideas. It is also encouraging that in recent years research in biology, physiology, and physics has opened up many new venues of looking at the energetic nature of reality, both in the body and the environment. These branches of science are beginning to create a language that will eventually allow Western science to account for the Chinese concepts of *qi*, *Fengshui*, and proper timing, demystifying Daoist concepts and making them more widely accessible to the general public.

The most important new concepts emerging from this research are measurable biomagnetic fields and bioelectricity (see Oschman, 2000). Biomagnetic fields are human energy centers that vibrate at different frequencies, storing and giving off energies not unlike the inner organs in the Chinese system. Their energetic output or vibrations can be measured, and it has been shown that the heart and the brain continuously pulse at extremely low frequencies (ELF). Through controlled measurements that biomagnetic fields are unbounded so that, for example, the field of the heart vibrates beyond the body and extends infinitely into space, verifying the Chinese conviction that people and the universe interact continuously on an energetic level.

Similarly, bioelectricity manifests in energy currents that crisscross the human body and are similar to the *qi*-channels in the Chinese system. Separate from and, in evolutionary terms, more ancient than the nervous system, these currents work through the so-called cytoskeleton, a complex net of connective tissue that is a continuous and dynamic molecular webwork. Also known as the "living matrix," this webwork contains so-called integrins or trans-membrane linking molecules which have no boundaries but are intricately interconnected. When touching the skin or inserting an acupuncture needle, the integrins make contact with all parts of the body through the matrix webwork. Based on this technique, wholeness is becoming an accepted concept, which sees "the body as an integrated, coordinated, successful system" and accepts that "no parts or properties are uncorrelated but all are demonstrably linked" (Oschman, 2000, p. 49, citing E. F. Adolph). It is now understood, that

> the electrical and electronic signals within the living system are honed to perfection. Each molecule has its electromagnetic signature. Each molecular interaction and each physiological process generates fields as well. All of these fields are compatible. In normal tissues, interferences and incompatibilities do not occur. Within the body, thousands of physiological and biochemical and electronic processes take place each second. (Oschman, 2000, p. 189)

More concretely, this means that each among the billions of cells in the human body, although it has a hundred thousand or more genes, knows that it has one, or sometimes a few, specific jobs to perform. More than that, it knows when to perform this job and when to return to dormancy or remain inactive. For example, as Judylaine Fine notes, "a liver cell might be concerned with producing proteins that can purify blood. When that liver cell is ready to purify blood, then the rest of its genes had better remain passive" (Fine, 1986, p. 10). Human life on the molecular biological level consists of the complex, integrated interaction of billions of entities that move extremely fast and smoothly, and in perfect coordination. Even more amazing, this scientifically established reality of human biology matches the vision of Daoists regarding the functioning of the human body as well as that of the universe at large.

This coordination is also expressed in the precise timing of activities in the body and of the body. It is well known that lunar phases and seasonal changes can have a massive effect on people's well-being, and what the Chinese describe as periods of life-*qi* and death-*qi* in the course of the day easily matches the biologically measured changes in blood pressure and sleepiness as people pass through a twenty-four hour period in the so-called circadian rhythm.

> Blood pressure is normally highest about 9:00 in the morning, staying at that level until mid-afternoon. At about 3:00 p.m. it starts to fall, reaching a low point around 1:00 a.m. It remains low for a few hours, until just before awakening. Around 5:00 a.m. it climbs sharply back toward its 9:00 a.m. peak. Then the whole cycle repeats itself. (Friedewald, 1995, p. 8)

Similarly, circadian sleep cycles create points of higher wakefulness or alerting in the early morning and early evening and low energy phases around midnight and in the early afternoon (Dement, 1999, p. 235). They are set before birth and remain strong even if people are isolated from the natural rhythm of light and darkness, and continuously following the natural pattern of the rise and fall of *qi*.

Not only with regard to the overall integrity of the human body and its matching of natural timing, modern scientists are coming to match Daoist views in their understanding that mental and emotional states are not separate from physical realities. They have begun to see intention as a kind of directed vibration that can have a disturbing or enhancing effect on the individual, others, and the environment. Mental attitudes, like the emotions in Daoism, are now understood to give rise to specific patterns of energy so that magnetic activity in the nervous system of the individual can spread through his or her body into outside energy fields. This understanding accounts for the efficacy of therapeutic touch and distant energy healing, during which the practitioner goes into a meditative state of mind and directs healing thoughts toward the patient (see Gerber, 1988). It also accounts for the sense of gloom or enthusiasm that can be found on the planet.

The living matrix as seen by science today is simultaneously a mechanical, vibrational, energetic, electronic, photonic, and informational network. It consists of a complex, linked pattern of pathways and molecules that forms a tensegrity system. A term taken originally from architecture where it is used in the structural description of domes, tents, sailing vessels, and cranes, tensegrity indicates a continuous tensional network (tendons) connected by a set of discontinuous elements (struts), which can also be fruitfully applied to the description of the wholeness of the body.

> The body as a whole, and the spine in particular, can usefully be described as tensegrity systems. In the body, bones act as discontinuous compression elements and the muscles, tendons and ligaments act as a continuous tensional system. Together the bones and tensional elements permit the body to change shape, move about, and lift objects. (Oschman, 2000, p. 153)

This understanding of the body as a tensegrity system allows for the analysis of physical and movement therapies, such as Qigong and Taiji quan and the Western methods of Feldenkrais, Alexander, and structural integration (Rolfing). It is becoming evident that bodily posture and modes of movement are essential to the way people feel and act. Straighter posture and better balance allow bioelectric energies to flow more freely, improve the ability of the body to adapt to changing situations (plasticity), and enhance the efficiency of the body's use. Also, correct movements may contribute to the release of traumas and tensions stored in the joints and muscles of the body, providing access to greater health and well-being.

But not only the body, the entire planet with all its life forms is in itself a tensegrity system, a living organism. In many ways replication the *qi*-notions in *Fengshui*, Western science has expressed this notion in the Gaia theory. First formulated by James Lovelock in 1979, it is based on the observation that the earth has a tendency to optimize conditions for terrestrial life. In a history of 4.5 eons or 1,000 million years, it should statistically have developed climatic changes toward extreme temperatures. However, this has never happened. The earth always kept things temperate, which shows that it tends to optimize the varying conditions by developing new agents that produce gases or chemicals counteracting problematic situations. Gaia as the living earth can be understood as the sum total of different individual modifications in planetary development, as the total interconnected network of all species. This is an ongoing thing. The earth continues to adapt and modify in accordance with different developments and to match its patterns to the network of species (Lovelock, 1979).

Gaia, like other living organisms, has vital organs at its core and redundant organs on its periphery. It is much easier for the earth to alter things on its outside but it resists changes in things deep inside. For example, during the ice age all above and below 45 degrees latitude froze over, which was about thirty percent of the earth's surface. Even with this amount of freezing, the planet still functioned, since its core in the center remained untouched. It is therefore less dangerous to alter climates and physical conditions in the far north or south of the planet and more hazardous to modify things around the equator. The cutting down of the rain forests is likely to have a greater impact on global conditions that any changes affected around the poles (see Thompson, 1987).

The living planet, like the body, is a completely harmonious and integrated system that functions ideally if treated properly and with care. It, too, has vastly different energetic patterns that all collaborate to make it work in its entirety and which are known to scientists as forms of geomagnetism. Like *qi*, geomagnetic waves can be smooth and harmonious or unpredictable and pathogenic – in each case creating an impact on the behavior of plants, animals, and humans. The Swedish chemist Svante Arrhenius, through extensive studies of this phenomenon, found that "biological rhythms correlate with rhythms and tides in the cosmic forces that surround the earth, that the electrical tension in the air influences biochemical reactions and thereby affects all living things," explaining weather sensitivity, changes in overall tension during periods of full moon or thunderstorms, allergies and other potential difficulties (Oschman, 2000, pp. 195-196).

Geomagnetic waves stay between the earth's surface and the ionosphere, which forms a resonant cavity in which waves are constantly going out and reflected back, interlinking all different aspects of planetary life in one integrated network. This vision is strongly reminiscent of the understanding of the close communication and interaction of Heaven, Earth, and Humanity in traditional Daoist thought. In both cases, no action or mentation is ever isolated or without resonance or consequence, micropulsations continuing to create states of harmony or imbalance. In some areas, these pulsations can lead places of great healing and cosmic integration, such as the mountains in Daoism, or to geopathic stress, to earth radiations that create particularly pathogenic sites where people feel uneasy or get easily sick (Oschman, 2000, p. 187).

To remedy this, modern science resorts less to *Fengshui* and measures of natural adaptation but instead creates mechanical devices, such as personal electromagnetic shielding devices that emit scalar waves and purportedly stabilize a person's brain waves at a frequency that is considered safe and beneficial (Oschman, 2000, p. 208). Traditional cultures, on the other hand, have tended to transform pathogenic sites through infusions of beneficial and positive consciousness, creating, as Itchak Bentov points out, nature spirits in the process. According to him, all natural sites have some dim consciousness of being, which is enhanced through the interaction with living creatures. "If an animal finds a hiding place in a rock formation," for example, "it will be grateful to the rock for shelter, and the rock will feel it," getting an ego-boost in the process (Bentov, 1977, p. 73). If more of this happens and especially if the site comes into contact with humans, this budding ego will evolve into the spirit of the rock which may even be able to make unusual things happen, creating a particular form of conscious atmosphere and geomagnetic vibration in its environment. When Daoists enter the mountains, they accordingly take great care lest they encounter some unevolved entities, using ritual and magical means to transform negative tendencies. They acknowledge the spiritual and conscious nature of all existence, striving to create harmony among all.

This attitude and effort stand in dire contrast to the tendencies of modern people, who in their interaction with the environment behave like cancer cells in the body. They become too big for their natural position; they do not listen to the naturally embedded commands to be active or latent; they see the production of more and more as their highest good; they go beyond the roles they are best suited for; and they try to take on and conquer everything. Becoming more and more aggressive, they conquer healthy aspects of the world just like cancerous growths attack healthy tissues. They monopolize nutrients and eventually become so large that they begin to oppress others around them. Like a tumor that takes over an organ, modern societies in their attitudes toward nature are ruthless and aggressive. Without considering the consequences in the larger balance of things, they turn night into day and day into night, transform wilderness into civilization, and disregard all ethical considerations. Like cancer cells, even modern people were once normal and had the perfect ability to sense what is right in the larger scheme of things. However, again like cancer cells, their attitudes' "deadliness comes from the fact that they spread and, as they do so, they interfere with so many of the normal functions of the body [and the world] that eventually nothing works at all" (Fine, 1986, p. 14).

Looking back to ancient Daoist wisdom and its subtle understanding of the integrated nature of healing and the earth may help raise awareness of just how much imbalance the planet and our bodies are already exposed to and offer some methods on how to begin the long process of overcoming and recovery.

## WORKS CITED

Bentov, Itchak. (1977) *Stalking the Wild Pendulum: On the Mechanics of Consciousness*. New York: E. P. Dutton.

Bokenkamp, Stephen. (1997) *Early Daoist Scriptures*, with a contribution by Peter Nickerson. Berkeley: University of California Press.

Campany, Robert F. (2001) "Ingesting the Marvelous: The Practitioner's Relationship to Nature According to Ge Hong." In: Norman Girardot, James Miller, and Liu Xiaogan (eds.), *Daoism and Ecology: Ways Within a Cosmic Landscape*, pp. 125□148. Cambridge, Mass.: Harvard University Press, Center for the Study of World Religions.

Dement, William C. (1999) *The Promise of Sleep*. New York: Dell.

Eichhorn, Werner. (1957) "T'ai-p'ing und T'ai-p'ing Religion," *Mitteilungen des Instituts für Orientforschung*, 5, pp. 113□140.

Field, Stephen L. (2001) "In Search of Dragons: The Folk Ecology of Fengshui." In: Norman Girardot, James Miller, and Liu Xiaogan (eds.), *Daoism and Ecology: Ways Within a Cosmic Landscape*, pp. 185□200. Cambridge, Mass.: Harvard University Press, Center for the Study of World Religions.

Fine, Judylaine. (1986) *Afraid to Ask: A Book for Families to Share about Cancer*. New York: Lothrop, Lee & Shepard Books.

Friedewald, Vincent E. (1995) *Ask the Doctor: Hypertension*. Kansas City: Andrews and McMeel.

Gerber, Richard. (1988) *Vibrational Medicine: New Choices for Healing Ourselves*. Santa Fe, N.M.: Bear and Company.

Hahn, Thomas H. (1988) "The Standard Taoist Mountain and Related Features of Religious Geography," *Cahiers d'Extrême-Asie*, 4, pp. 145□156.

________. (2000) "Daoist Sacred Sites." In: Livia Kohn (ed.), *Daoism Handbook*, pp. 683□707. Leiden; Boston: Brill.

Hendrischke, Barbara, and Benjamin Penny. (1996) "The 180 Precepts Spoken by Lord Lao: A Translation and Textual Study," *Taoist Resources*, 6:2, pp. 17□29.

Henricks, Robert. (2000) *Lao Tzu's Tao Te Ching: A Translation of the Startling New Documents Found at Guodian*. New York: Columbia University Press.

Inoue, Yutaka. (1992) "Rokutei, Rokuka jin no henyu," *Tōhō shūkyō*, 80, pp. 15□32.

Kaltenmark, Max. (1953) *Le Lie-sien tchouan*. Peking: Université de Paris, Publications du Centre d'Etudes Sinologiques de Pékin.

Kaptchuk, Ted J. (1983) *The Web that Has No Weaver: Understanding Chinese Medicine*. New York: Congdon & Weed.

Kohn, Livia. (2001) *Daoism and Chinese Culture*. Cambridge, Mass.: Three Pines Press.

________. (2004) *Cosmos and Community: The Ethical Dimension of Daoism*. Cambridge, Mass.: Three Pines Press.

________. (2005) *Health and Long Life: The Chinese Way*. Cambridge, Mass.: Three Pines Press.

Komjathy, Louis. (2002) *Title Index to Daoist Collections*. Cambridge, Mass.: Three Pines Press.

Legge, James. (trans.) (1966) "*Li Ki*," *The Sacred Books of China: The Texts of Confucianism*, Parts 3□4. Delhi: Motilal Banarsidass.

Lip, Evelyn. (1995) *Feng Shui: Environments of Power: A Study of Chinese Architecture*. London: Academy Editions.

Liu, Xiaogan. (1991) "Wuwei (Non-Action): From *Laozi* to *Huainanzi*," *Taoist Resources*, 3:1, pp. 41□56.

________. (1998) "Naturalness (Tzu-jan), the Core Value in Taoism: Its Ancient Meaning and Its Significance Today." In: Livia Kohn and Michael LaFargue (eds.), *Lao-tzu and the Tao-te-ching*, pp. 211□228. Albany, N.Y.: State University of New York Press.

Liu, Zhengcai. (1990) *The Mystery of Longevity*. Beijing: Foreign Languages Press.

Lovelock, James. (1979) *Gaia: A New Look at Life on Earth*. New York: Oxford University Press.

Major, John S. (1993) *Heaven and Earth in Early Han Thought: Chapters Three, Four, and Five of the Huainanzi*. Albany: State University of New York Press.

Malek, Roman. (1985) *Das Chai-chieh lu: Materialien zur Liturgie im Taoismus*. Frankfurt am Main: P. Lang.

Miura, Kunio. (1983) "Doten fukuchi ron," *Tōhō shūkyō*, 61, pp. 1□23.

Mugitani, Kunio. (1987) *Yōsei Enmeiroku Kunchi*. Tokyo: Mombushō.

Oschman, James L. (2000) *Energy Medicine: The Scientific Basis*. Edinburgh: Churchill Livingstone.

Penny, Benjamin. (1996) "Buddhism and Daoism in *The 180 Precepts Spoken by Lord Lao*," *Taoist Resources*, 6:2, pp. 1□16.

Porter, Bill. (1993) *The Road to Heaven: Encounters with Chinese Hermits*. San Francisco: Mercury House.

Robinet, Isabelle. (1989) "Visualization and Ecstatic Flight in Shangqing Taoism." In: Livia Kohn (ed.), *Taoist Meditation and Longevity Techniques*, pp. 157□190. Ann Arbor: Center for Chinese Studies, University of Michigan.

________. (1993) *Taoist Meditation: The Mao-shan Tradition of Great Purity*, trans., Julian F. Pas and Norman J. Girardot. Albany: State University of New York Press.

Rossbach, Sarah. (1983) *Feng Shui: The Chinese Art of Placement*. New York: Dutton.

________ and Lin, Yun. (1994) *Living Color: Master Lin Yun's Guide to Feng Shui and the Art of Color*. New York: Kodansha International.

Schipper, Kristofer M. (1975) *Concordance du Tao Tsang: Titres des ouvrages*. Paris: Publications de l'Ecole Française d'Extrême-Orient.

________. (2001) "Daoist Ecology: The Inner Transformation. A Study of the Precepts of the Early Daoist Ecclesia." In: Norman Girardot, James Miller, and Liu Xiaogan (eds.), *Daoism and Ecology: Ways Within a Cosmic Landscape*, pp. 79□94. Cambridge, Mass.: Center for the Study of World Religions, Harvard Divinity School: Harvard University Press.

Schwartz, Benjamin I. (1985) *The World of Thought in Ancient China.* Cambridge, Mass.: Harvard University Press.

Switkin, Walter. (1987) *Immortality: A Taoist Text of Macrobiotics.* San Francisco: H. S. Dakin Company.

Taoist Association of China. (2002) *Taoism.* Beijing: Foreign Languages Press.

Thompson, William Irvin. (ed.) (1987) *Gaia, A Way of Knowing: Political Implications of the New Biology.* Hudson, N.Y.: Lindisfarne Press.

Ware, James R. (1966) *Alchemy, Medicine and Religion in the China of AD 320. The Nei P'ien of Ko Hung.* Cambridge, Mass.: MIT Press.

# ABOUT THE CONTRIBUTORS

Anis Ahmad, Vice-Chancellor, Riphah International University, Pakistan.

Gerhold K. Becker, Visiting Professor, Graduate School of Philosophy, Assumption University of Thailand, Thailand; Chair Professor of Philosophy (ret.), and Founding Director of Centre for Applied Ethics, Hong Kong Baptist University, Hong Kong Special Administrative Region, China.

Jonathan Chan, Associate Professor, Department of Religion and Philosophy, and Associate Director of Centre for Applied Ethics, Hong Kong Baptist University, Hong Kong Special Administrative Region, China.

Robert Elliot, Pro Vice-Chancellor (International and Development), and Professor of Philosophy, University of the Sunshine Coast, Australia.

King-Tak Ip, Adjunct Professor, Department of Religion and Philosophy, and Chaplain, Hong Kong Baptist University, Hong Kong Special Administrative Region, China.

Livia Kohn, Professor, Department of Religion, Boston University, U.S.A.

Y. S. Lo, Lecturer, Department of Philosophy, La Trobe University, Australia.

Michael S. Northcott, Professor of Ethics, School of Divinity, University of Edinburgh, The United Kingdom.

Ingmar Persson, Professor, Department of Philosophy, Göteborg University, Sweden.

Pragati Sahni, Lecturer, Department of Philosophy, University of Delhi, India.

# INDEX

**VIBS**

The **Value Inquiry Book Series** is co-sponsored by:

Adler School of Professional Psychology
American Indian Philosophy Association
American Maritain Association
American Society for Value Inquiry
Association for Process Philosophy of Education
Canadian Society for Philosophical Practice
Center for Bioethics, University of Turku
Center for Professional and Applied Ethics, University of North Carolina at Charlotte
Central European Pragmatist Forum
Centre for Applied Ethics, Hong Kong Baptist University
Centre for Cultural Research, Aarhus University
Centre for Professional Ethics, University of Central Lancashire
Centre for the Study of Philosophy and Religion, University College of Cape Breton
Centro de Estudos em Filosofia Americana, Brazil
College of Education and Allied Professions, Bowling Green State University
College of Liberal Arts, Rochester Institute of Technology
Concerned Philosophers for Peace
Conference of Philosophical Societies
Department of Moral and Social Philosophy, University of Helsinki
Gannon University
Gilson Society
Haitian Studies Association
Ikeda University
Institute of Philosophy of the High Council of Scientific Research, Spain
International Academy of Philosophy of the Principality of Liechtenstein
International Association of Bioethics
International Center for the Arts, Humanities, and Value Inquiry
International Society for Universal Dialogue
Natural Law Society
Philosophical Society of Finland
Philosophy Born of Struggle Association
Philosophy Seminar, University of Mainz
Pragmatism Archive at The Oklahoma State University
R.S. Hartman Institute for Formal and Applied Axiology
Research Institute, Lakeridge Health Corporation
Russian Philosophical Society
Society for Existential Analysis
Society for Iberian and Latin-American Thought
Society for the Philosophic Study of Genocide and the Holocaust
Unit for Research in Cognitive Neuroscience, Autonomous University of Barcelona
Whitehead Research Project
Yves R. Simon Institute

Titles Published

Volumes 1 - 167 see www.rodopi.nl

168. Michael W. Riley, *Plato's* Cratylus: *Argument, Form, and Structure*. A volume in **Studies in the History of Western Philosophy**

169. Leon Pomeroy, *The New Science of Axiological Psychology*. Edited by Rem B. Edwards. A volume in **Hartman Institute Axiology Studies**

170. Eric Wolf Fried, *Inwardness and Morality*

171. Sami Pihlstrom, *Pragmatic Moral Realism: A Transcendental Defense.* A volume in Studies in **Pragmatism and Values**

172. Charles C. Hinkley II, *Moral Conflicts of Organ Retrieval: A Case for Constructive Pluralism.* A volume in **Values in Bioethics**

173. Gábor Forrai and George Kampis, Editors, *Intentionality: Past and Future.* A volume in **Cognitive Science**

174. Dixie Lee Harris, *Encounters in My Travels: Thoughts Along the Way*. A volume in **Lived Values:Valued Lives**

175. Lynda Burns, Editor, *Feminist Alliances*. A volume in **Philosophy and Women**

176. George Allan and Malcolm D. Evans, *A Different Three Rs for Education*. A volume in **Philosophy of Education**

177. Robert A. Delfino, Editor, *What are We to Understand Gracia to Mean?*: *Realist Challenges to Metaphysical Neutralism*. A volume in **Gilson Studies**

178. Constantin V. Ponomareff and Kenneth A. Bryson, *The Curve of the Sacred*: *An Exploration of Human Spirituality*. A volume in **Philosophy and Religion**

179. John Ryder, Gert Rüdiger Wegmarshaus, Editors, *Education for a Democratic Society: Central European Pragmatist Forum*, *Volume Three*. A volume in **Studies in Pragmatism and Values**

180. Florencia Luna, *Bioethics and Vulnerability: A Latin American View*. A volume in **Values in Bioethics**

181. John Kultgen and Mary Lenzi, Editors, *Problems for Demo*cracy. A volume in **Philosophy of Peace**

182. David Boersema and Katy Gray Brown, Editors, *Spiritual and Political Dimensions of Nonviolence and Peace*. A volume in **Philosophy of Peace**

183. Daniel P. Thero, *Understanding Moral Weakness*. A volume in **Studies in the History of Western Philosophy**

184. Scott Gelfand and John R. Shook, Editors, *Ectogenesis: Artificial Womb Technology and the Future of Human Reproduction*. A volume in **Values in Bioethics**

185. Piotr Jaroszyński, *Science in Culture*. A volume in **Gilson Studies**

186. Matti Häyry, Tuija Takala, Peter Herissone-Kelly, Editors, *Ethics in Biomedical Research: International Perspectives*. A volume in **Values in Bioethics**

187. Michael Krausz, *Interpretation and Transformation: Explorations in Art and the Self*. A volume in **Interpretation and Translation**

188. Gail M. Presbey, Editor, *Philosophical Perspectives on the "War on Terrorism."* A volume in **Philosophy of Peace**

189. María Luisa Femenías, Amy A. Oliver, Editors, *Feminist Philosophy in Latin America and Spain*. A volume in **Philosophy in Latin America**

190. Oscar Vilarroya and Francesc Forn I Argimon, Editors, *Social Brain Matters: Stances on the Neurobiology of Social Cognition*. A volume in **Cognitive Science**

191. Eugenio Garin, *History of Italian Philosophy*. Translated from Italian and Edited by Giorgio Pinton. A volume in **Values in Italian Philosophy**

192. Michael Taylor, Helmut Schreier, and Paulo Ghiraldelli, Jr., Editors, *Pragmatism, Education, and Children: International Philosophical Perspectives*. A volume in **Pragmatism and Values**

193. Brendan Sweetman, *The Vision of Gabriel Marcel: Epistemology, Human Person, the Transcendent*. A volume in **Philosophy and Religion**

194. Danielle Poe and Eddy Souffrant, Editors, *Parceling the Globe: Philosophical Explorations in Globalization, Global Behavior, and Peace.* A volume in **Philosophy of Peace**

195. Josef Šmajs, *Evolutionary Ontology: Reclaiming the Value of Nature by Transforming Culture*. A volume in **Central-European Value Studies**

196. Giuseppe Vicari, *Beyond Conceptual Dualism: Ontology of Consciousness, Mental Causation, and Holism in John R. Searle's Philosophy of Mind.* A volume in **Cognitive Science**

197. Avi Sagi, *Tradition vs. Traditionalism: Contemporary Perspectives in Jewish Thought*. Translated from Hebrew by Batya Stein. A volume in **Philosophy and Religion**

198. Randall E. Osborne and Paul Kriese, Editors, *Global Community: Global Security.* A volume in **Studies in Jurisprudence**

199. Craig Clifford, *Learned Ignorance in the Medicine Bow Mountains: A Reflection on Intellectual Prejudice.* A volume in **Lived Values: Valued Lives**

200. Mark Letteri, *Heidegger and the Question of Psychology: Zollikon and Beyond.* A volume in **Philosophy and Psychology**

201. Carmen R. Lugo-Lugo and Mary K. Bloodsworth-Lugo, Editors, *A New Kind of Containment: "The War on Terror," Race, and Sexuality.* A volume in **Philosophy of Peace**

202. Amihud Gilead, *Necessity and Truthful Fictions: Panenmentalist Observations*. A volume in **Philosophy and Psychology**

203. Fernand Vial, *The Unconscious in Philosophy, and French and European Literature: Nineteenth and Early Twentieth Century.* A volume in **Philosophy and Psychology**

204. Adam C. Scarfe, Editor, *The Adventure of Education: Process Philosophers on Learning, Teaching, and Research*. A volume in **Philosophy of Education**

205. King-Tak Ip, Editor, *Environmental Ethics: Intercultural Perspectives.* A volume in **Studies in Applied Ethics**